Jennings, D. L.

Raspberries and blackberries

$39.95

DATE			

Raspberries and Blackberries:
Their Breeding, Diseases and Growth

APPLIED BOTANY AND CROP SCIENCE

SERIES EDITORS: R. W. Snaydon, University of Reading, England, J. M. Barnes, Co-operative State Research Service, United States Department of Agriculture, Washington, USA, and F. L. Milthorpe†, Macquarie University, New South Wales, Australia

Physiological Ecology of Forest Production *J. J. Landsberg*
Weed Control Economics *B. A. Auld, K. M. Menz and C. A. Tisdell*
Improving Vegetatively Propagated Crops *A. J. Abbott and R. K. Atkin*
Raspberries and Blackberries: Their Breeding, Diseases and Growth *D. L. Jennings*

† Deceased

Raspberries and Blackberries: Their Breeding, Diseases and Growth

D. L. JENNINGS

Scottish Crop Research Institute,
Invergowrie,
Scotland

1988

ACADEMIC PRESS
Harcourt Brace Jovanovich, Publishers
London San Diego New York
Boston Sydney Tokyo Toronto

ACADEMIC PRESS LIMITED
24/28 Oval Road,
London NW1 7DX

United States Edition published by
ACADEMIC PRESS INC.
San Diego, CA 92101

British Library Cataloguing in Publication Data

Jennings, D. L.
Raspberries and blackberries : their breeding, diseases and growth.
1. Raspberries 2. Blackberries
I. Title
634′.711 SB386.R3

ISBN 0-12-384240-9

Photoset in Great Britain by
Rowland Phototypesetting Limited, Bury St Edmunds, Suffolk
and printed by St Edmundsbury Press Limited, Bury St Edmunds, Suffolk

Preface

Raspberries and blackberries have captivated the interest of botanists and gardeners for centuries. It was the medicinal properties of raspberries that first attracted the attention of herbalists, but this was soon followed by the concerted efforts of plant breeders, both amateur and non-amateur, to produce bigger, better and more flavoursome fruits. More recently, breeders have largely restructured the plant and succeeded in breeding cultivars able to crop over a longer season or in new environments, or to resist or tolerate many of the diseases and hazards that affect the crop. In blackberries the vast reservoir of variation first prompted taxonomists to describe a multitude of taxonomic forms, and this was followed by an equal determination by cytogeneticists to unravel the mechanisms which allow the breeding system to create and maintain them. As with the raspberry, plant breeders have now bred greatly improved cultivars that are adapted to diverse situations. Moreover, the extremely diverse nature of the genus *Rubus* has provided opportunities for the breeder to create new kinds of fruit by crossing among distantly related species within the genus. Hence fruits like the Loganberry, Boysenberry and Tayberry have appeared in comparatively recent times.

As the economic importance of raspberries and blackberries has increased, plant pathologists and physiologists have greatly added to our understanding of the crops and provided an increasing fund of knowledge. It is therefore timely to review the past and bring together some of the knowledge that has been gained from research. I have not attempted to cover the literature comprehensively, but trust that my selection provides suitable entry points for a more complete coverage where this is required. Parts of the text are inevitably technical, but I hope that the book will interest enthusiastic growers as well as scientists, both in the non-technical historical parts and in the scientific parts, which I have tried to present with a minimum of technical detail and jargon.

Several of my friends and colleagues commented on the drafts of various chapters: in particular, I thank most sincerely Teifion Jones, Brian Williamson, Stuart Gordon and David Trudgill of the Scottish Crop Research Institute. They cannot be held responsible for errors or omissions. I also thank Maureen Murray and my wife Joan, who shared the typing, Eleanor Brydon, who helped with the references, the Scottish Crop Research

Institute for allowing me to use official photographs and Tom Geoghegan, Stuart Malecki and Gregor Menzies who took the photographs and prepared the prints.

D. L. Jennings

Contents

9 Diseases Caused by Pests

10 Diseases Caused by Nematodes

11 The Growth Cycle

12 Stems, Leaves and Roots

1 Introduction: the Genus *Rubus*

Rubus is one of the most diverse genera in the plant kingdom. It contains a wide spectrum of wild species in addition to those from which domesticated cultivars of edible fruits have been selected. Focke in his three monographs published from 1910 to 1914 divided the genus into 12 subgenera; of these, one contains raspberries, another contains blackberries, two contain arctic fruits and one contains ornamentals. The remainder contain few, if any, domesticated species. The taxonomy and distribution of *Rubus* species was still very imperfectly known at the time of Focke's work, even though the collecting tours at the early part of the century had gathered a great quantity of new material from Asia, Africa and South America. Knowledge of the species is now more complete, but many workers prefer to follow Focke's basic subdivision of the genus into 12 subgenera, which is as follows:

Subgenus 1	Chamaemorus	1 species	(R. *chamaemorus*, the cloudberry, see p. 68)
Subgenus 2	Dalibarda	5 species	No pomological value
Subgenus 3	Chamaebatus	5 species	No pomological value
Subgenus 4	Comaropsis	2 species	No pomological value
Subgenus 5	Cylactis	14 species	(Arctic berries, see p. 68–71)
Subgenus 6	Orobatus	19 species	
Subgenus 7	Dalibardastrum	4 species	No pomological value (Asia)
Subgenus 8	Malachobatus	114 species	No pomological value Some ornamentals
Subgenus 9	Anoplobatus	6 species	(Flowering raspberries)
Subgenus 10	Idaeobatus	200 species	(Raspberries, see p. 3–37)
Subgenus 11	Lampobatus	10 species	No pomological value
Subgenus 12	Eubatus	Very large number of species (Blackberries, see p. 39–57)	

The first five of these subgenera contain perennial herbaceous plants with creeping stems or a creeping rootstock bearing annual erect flowering shoots. The others contain shrubs with stems which persist for at least two years. There is considerable differentiation of species in the *Eubatus* (blackberries), *Idaeobatus* (raspberries) and *Malachobatus* subgenera. The latter, which contains no important fruit-bearing species, occurs principally in south-eastern Asia, Japan, Malaya, Australia and Madagascar, and is absent from more temperate zones and from areas bordering the Atlantic. Focke considered that the subgenera *Orobatus* and *Chamaebatus* were allied to it.

The *Idaeobatus* subgenus has a northerly distribution, principally Asia, but also in east and south Africa, Europe and North America. It has some 200 species, of which the most important are the European red raspberry (*Rubus idaeus* subsp. *vulgatus* Arrhen.), the North American red raspberry (*R. idaeus* subsp. *strigosus* Michx) and the black raspberry (*R. occidentalis L.*). Purple raspberries are hybrids of red and black raspberries, though they were given the specific rank of *R. neglectus* Peck at one time. Many other species of this subgenus have been used by plant breeders as donors of improvements in the breeding of red raspberries, mostly in respect of resistance to diseases and pests.

The subgenus *Cylactis* is close to the *Idaeobatus*. It contains the important arctic raspberries *R. arcticus* L and *R. stellatus* Sm.

The subgenus *Eubatus* is extremely variable and complex. It contains all the blackberries and dewberries and has several sections in South America, a very prominent one in Europe and another in North America. It is virtually absent from Pacific- and Indian-ocean regions. The remarkable combination of sexual and subsexual reproduction in the *Moriferi* section has given rise to thousands of taxonomic units which have been given specific rank, though many of them are interfertile and it is often not possible to assign cultivars to individual species. The species of the subgenus range from evergreen subtropical species to deciduous ones adapted to northern Canada, and the ploidy range extends from diploid ($2x = 2n = 14$) to dodecaploid ($12x = 2n = 84$). The octoploid species of the *Ursini* section from western North America combine well with raspberries to give hybrids such as the Loganberry and Tayberry that have achieved major economic importance.

Further study may well show that some of the small subgenera contain hybrids between species of diverse subgenera: Darrow (1955a,b), for example, speculated that *R. roseus* of the *Orobatus* subgenus may be a hybrid of *Eubatus* and *Anoplobatus* species. Possibly some of the small subgenera recognized by Focke consist only of hybrids between subgenera which are not readily associated with their parents because they combine characteristics of distant groups, or because their parental species are now extinct.

2 Red Raspberries

Raspberries belong to the subgenus *Idaeobatus*, whose species are distinguished by the ability of their mature fruits to separate from the receptacle. The subgenus occurs on all five continents and is particularly well represented in the northern hemisphere, being most diverse in temperate and subtropical regions of eastern Asia. Over 200 species are now recognized; several of them have been domesticated, but the red and black raspberries of Europe, Asia and North America are the only ones grown on a large scale. References on the history of their domestication are given by Bailey (1898), Bunyard (1922, 1925), Card (1898), Darrow (1937), Haskell (1954), Hedrick (1925) and Roach (1985).

2.1 RED RASPBERRIES IN THE WILD

Red raspberries are widely distributed in all temperate regions of Europe, Asia and North America. They are variable, but two well defined types are readily distinguished. Focke (1911) regarded them as subspecies and used the designations *R. idaeus* subsp. *vulgatus* Arrhen. and *R. idaeus* subsp. *strigosus* Michx for the two forms. Rozanova (1939b) considered that the morphological differences and specific geographical distributions of the two forms justified giving them specific rank, though she recognized that they were both ecotypes of *R. idaeus* L. *R. idaeus vulgatus* Arrhen. is the diploid European form with glandless inflorescences and thimble-shaped fruits, which extends from near the polar circle to the mountains of the Caucasus in Asia minor, and *R. idaeus strigosus* (Michx.) is the diploid form of North America and East Asia with glandular inflorescences and round fruits. The two forms intercross readily and their hybrids show little or no sterility. They are referred to here as *R. idaeus* and *R. strigosus*. Other forms include *R. idaeus melanolasius* Focke, from north-west America, and three tetraploids from eastern Asia: *R. idaeus melanolasius* (Michx.) Maxim, *R. idaeus sibiricus* Komarov and *R. idaeus sachalinensis* Leveille. Rozanova (1939a,b) considered that these forms were autotetraploid and sufficiently similar morphologically to be included in one species derived from *R. strigosus* and designated *R. sachalinensis* Leveille. She later recognized a frost-resistant

Caucasian form with large fruit and named it *R. vulgatus buschii* (Rozanova, 1945).

In contrast to blackberries, where cultivated forms are often indistinguishable from wild ones, cultivated forms of raspberries are very different from their wild relatives. The difference in cane production is particularly large; Haskell (1960), for example, reported that wild forms in their second year produced about 70 canes per plant within a range of 30 to 130, compared to less than 20 per plant in cultivated forms. In fact cultivated forms often do produce about 70 canes early in the season, but the number becomes drastically reduced later, because they do not all continue to develop. Individual canes in wild forms are shorter and thinner than in cultivated forms, and they have short thin laterals bearing small flowers. The contrast in fruit size is particularly marked: Haskell found that the fruit size of cultivated raspberries was two to three times as large as that of the wild forms. Both the number and size of individual drupelets in wild raspberries are small, and the fruit are generally soft and often crumbly because of poor cohesion between the drupelets; they are usually easy to pick, but are often hidden by vigorous cane growth. Haskell considered the flavour of wild forms to be inferior too, but Rousi (1965) selected wild forms in Finland which were thought superior in flavour. Mišić and Tešović (1973) had similar success in the mountains of Yugoslavia, where they collected wild raspberries with large, tasty and aromatic fruit which were valuable for breeding.

Although Haskell's material was collected from widely separated regions of Britain, he found no association between geographical and ecological factors in any of the characters that he studied. Similarly, Rousi could not distinguish ecotypes or geographical races amongst his Finnish populations, though he found that some characters showed a differentiation in a north–south direction, that others did so in an east–west direction, and that there was a tendency for certain character combinations to occur in certain areas and in ecologically similar conditions. Rozanova (1939a) reported that populations from north and north-east Europe generally had a more spreading habit, hardier canes and flowered earlier than those from further south, which tended to be late flowering and included both spreading and erect forms. Jennings (1964a) showed that populations in Scotland were differentiated from each other even when separated by quite small distances, though they did not form distinct ecotypes. He found that plants from exposed sites tended to have relatively short canes and relatively late bud-burst and maintained these characteristics when grown in a favourable environment; there was a significant correlation between cane height and earliness of bud-burst in the latter environment. Keep (1972), in a survey of wild raspberries from Europe and Britain, found that parents of increasing altitudinal origin tended to give progenies with increasing resistance to mildew; thus the total percentage of resistance of all progenies was 6.6, 59.6, 64.3 and 97.2 for parental altitudes of less than 500 m, 500–1000 m, 1000–1500 m and over 1500 m, respectively. She suggested that mildew resistance may be a secondary effect of adaptation to the harsh climatic conditions of high altitude, since it seemed unlikely that selection pressure for mildew resistance would be greater at high altitudes, where the disease is usually less common. This is also found in other fruits (Keep, 1983).

The American red raspberry (*R. strigosus* Michx.) occurs widely in both mountain and lowland areas of North America. Its canes tend to be more hardy than those of *R. idaeus*, and usually thinner and more erect. Its fruit are usually round and seldom conical like those of *R. idaeus*. It is extremely variable in the wild, but the variations have attracted little study. Van Adrichem (1972) studied samples collected in British Columbia and northern Alberta and compared them with the Canadian cultivar "Trent", which has both *R. strigosus* and *R. idaeus* in its ancestry. Many of his wild collections had taller canes than that of the compared cultivar—in contrast with the results for *R. idaeus*. More surprisingly, some of the samples had significantly heavier fruit than "Trent" and some were not significantly different from it. Although the collection area contained considerable differences in climate and elevation, van Adrichem found no evidence of ecotype differentiation; flowering time was not related to location or to elevation, even though there was a 10-day spread between the earliest and latest. Dorsey (1921) made a study of wild raspberries in Manitoba, Canada and found variation in hardiness and morphological characteristics, while from studies in British Columbia, Daubeny and Stary (1982) discovered valuable genes for aphid resistance (see p. 134), and Vrain and Daubeny (1986) discovered resistance to *Pratylenchus penetrans* (see p. 142).

Many students of wild populations of *R. idaeus* have found polymorphism for hairy and subglabrous canes, and Jennings (1963a) commented that many cultivars that are believed to be closely related to wild ancestors are heterozygous for gene *H*, which determines this difference. The same is true, though to a lesser extent, of yellow and red fruit colour determined by gene *T* (Table 2.1). Wild raspberries are self-incompatible, and a possible explanation of this polymorphism is that gene *H* is closely linked to the incompatibility gene, however, Keep (1972) has shown that this is not so. A likely explanation for the polymorphism of both gene *H* and gene *T* is their association with a balanced lethal system, which selects heterozygous individuals and preserves recessive genes when their homozygous form is at a selective disadvantage (Jennings, 1967a). That the homozygous recessive forms of gene *T* are at a selective disadvantage was first noticed by Darwin, who wrote that yellow-fruited raspberries are not much touched by birds, the chief agents of their dissemination.

Table 2.1.

	No. that bred true for gene H	No. that segregated for genes H and h	No. that bred true for gene h
Wild raspberry samples	12	35	2
Hairy-caned raspberry cultivars*	0	5	—
	No. that bred true for gene T	**No. that segregated for genes T and t**	**No. that bred true for gene t**
Wild raspberry samples	35	13	1
Red raspberry cultivars*	7	16	—

*Includes only cultivars believed to be closely related to wild populations (see Jennings, 1963a).

2.2 THE DOMESTICATION OF THE RED RASPBERRY

Many writers have extolled the virtues of the people who lived over 2000 years ago in the city of Troy, in the foothills of Mount Ida in Asia minor. Their virtues included being the first to work iron and copper, and being responsible for introducing music and rhythm to Greece. Here they are mentioned for being the first to appreciate the delights of the raspberry.

Probably the first mention of raspberries in literature was by Cato, who wrote in pre-Christian times, but it was Pliny the Elder, writing about 45 A.D., who first described how the Greeks called them "ida" fruits, after Mount Ida and the people of that area who gathered them. It seems likely that the fruits were not cultivated at that time, but gathered from the wild. They certainly did not attain any importance, because they were not mentioned by Athenaeus, Theocritus or Virgil or by any of the other eminent writers of the day. The plant was more important as a medicine than as a food, as the blossom was used to make an eye ointment or a stomach draught. Dioscorides referred to them only in a medicinal context. However, by the fourth century A.D., Palladius, a Roman agricultural author, was writing about raspberries as a cultivated fruit, and raspberry seeds have been found at the sites of Roman forts in Britain. The name "ida" persisted, and was later used by Linnaeus to derive the specific name *idaeus* for the red raspberry; for the genus, he used the name *Rubus*, from the Latin Ruber, meaning red. However, although Mount Ida is generally believed to have inspired the name, modern botanists have failed to find raspberries growing there, and some have suggested that the Ide mountains of Turkey may be the correct place of origin.

We read that Henry III's clerk bought raspberries for the king's drinks, but nothing was written of them from these early times until almost the 16th century, when Matthioli wrote that in Bohemia they were being taken from woods to gardens, and Ruellius wrote that they were everywhere cultivated in gardens. Yellow forms were recorded by Canerarius in 1588 and by Clusius in 1601. Raspberries also figure in a German herbal of that time and in William Turner's *Names of Herbes* published in England in 1548. Turner mentioned that "*Rubus idaeus*, in English raspes or hyndberries, . . . grow most plenteously in the woddes in east Friesland. . . . They also grow in certayne gardines in Englande. . . . The taste of it is soure." Heresboch's book, published in 1570, also indicated that raspberries were not yet popular, but John Parkinson in his *Paradisi in Sole* published in 1629 wrote of red, white and thornless "raspis-berries" suitable for the English climate (see Fig. 2.1). In Shakespeare's day they were grown by the market gardeners of Chiswick and Brentford for London markets. Tradescant's catalogue of 1656 mentioned three cultivars and a fourth one referred to as purpureo, presumably because it had a darker colour. Worlidge in 1629 gave the first record of a variety with large fruit, and the first clue that gardeners were successfully selecting for types with improved fruits. Richard Weston mentioned a twice-bearing kind in 1780. By 1829 George Johnston in his

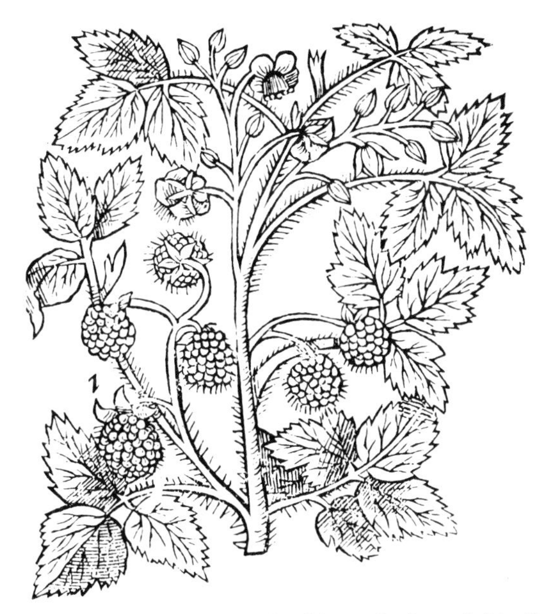

Fig. 2.1. "Raspis-berries" from John Parkinson's *Paradisi in Sole* published in 1629. (Courtesy of the John Innes Institute library.)

History of English Gardening could list 23 cultivated varieties, but this number was low compared with the number of varieties being described for other fruits. In America, W. Prince in 1771 was the first to sell raspberry plants commercially in New York, and his son, W. R. Prince, published a *Pomological Manual* in1832 with a description of 20 cultivars. By this time raspberries as a garden fruit had truly arrived on both continents.

Medicinal uses of raspberries continued to be common, and raspberry-leaf tea, one of the oldest of all herbal infusions, was given as an approved aid for taking during confinements in the most ancient of herbal books. Moreover, recent research has added scientific support by showing that the leaves contain a water-extractable principle capable of relaxing the uterine muscles of test animals (Burn and Withell, 1941). Garard Dewes, writing in 1578, recommended that "the flowers of raspis are good to be bruysed with hony and layde to the inflammations and hoate humours gathered togither in the eyes . . . for it quencheth such hoate burnings. They be also good to be drunken with water of them that have weake stomackes." Dewes' use of the word raspis is typical of writers in this period. Some earlier references use "raspa", but most 16th century writers use either "raspis" or the German word "hindberrie". Although it is commonly believed that the modern form "raspberry" refers to the fruit's sharp "rasping" flavour, it seems equally likely that it was derived from the old Anglo-Saxon word "Resp", meaning a shoot or sucker.

Clearly, the evolution of domesticated raspberries is a comparatively recent event, with a history of no more than 400–500 years, a very short period compared with that of most of our fruits.

2.3 EARLY RASPBERRY CULTIVARS IN BRITAIN

Raspberry culture in the 19th and early part of the 20th century was marked by the efforts of several enthusiastic gardeners and nurserymen who produced a series of cultivars of considerable merit. The outstanding cultivar at first was "Red Antwerp", which was notable for its large fruit. According to Brookshaw's *Pomona Britannica*, published in 1812, it was raised in about 1800 by a Mr. Cornwall of Barnet, who gave it its name because it grew as large as the "Yellow Antwerp". The latter cultivar had been received from the Governor of Antwerp at an earlier date and is thought to have originated in Hungary. "Fastolff" was discovered in about 1820 near an old castle formerly owned by Sir John Fastolff and was probably a seedling of "Red Antwerp". Bunyard's of Waldershare Gardens near Dover introduced another major success. Their "Superlative" was discovered near Dover by a Mr. Merryfield and introduced in 1888. Grubb (1922) thought that it was synonymous with "Northumberland Filbasket". It was considered the giant of the 1900s. John Baumforth of Pontefract released his famous "Baumforth" variety in about 1865, derived from "Northumberland Filbasket". Many cultivars came from the nurseries of Messrs. Laxtons (some examples are "Bountiful", "Prolific", "Renown", "Reward" and "Yellow Hammer") and Messrs. Veitch ("November Abundance", for example). Carters of Yorkshire introduced the important cultivars "Carter's Prolific" and "Semper Fidelis" in 1885. However, T. B. Pyne of Topsham Nurseries in Devon was undoubtedly the most successful nurseryman. He was reputed to have obtained new cultivars by transplanting self-sown seedlings; his successes included "Devon", "Red Cross", "Park Lane", "Imperial", "Mayfair", "Better Late" (a raspberry–blackberry hybrid), "Heytor" (autumn fruiting) and "Royal". "Park Lane" and its derivative "Mayfair" were said to be the best flavoured of all raspberries and "Devon" was noted for the firmness of its fruits, but the outstanding success was "Royal". "Pyne's Royal" was released in 1913 and rapidly gained a reputation for its high yield, large fruit and excellent flavour, almost equal to its near contemporary "Lloyd George", which was found by J. J. Kettle in a wood in Kent and distributed in 1919. Both "Pyne's Royal" and "Lloyd George" were later used extensively for controlled breeding at the East Malling Research Station.

Another valuable parent was "Norfolk Giant", which was found by a horticultural adviser in Norfolk and introduced in 1926. Up until this time new cultivars were obtained by chance discoveries or by growing families of open-pollinated seedlings. The first controlled breeding in Britain was not done until Grubb began his breeding work at East Malling Research Station. Grubb (1922) also made a study of all the cultivars being grown in Britain in the 1920s. This proved difficult, because most of the raspberry stocks that he obtained proved to be mixtures, or contained a high proportion of rogues; others were known by several synonymous names and, most common of all, distinct cultivars often occurred under one name. Grubb solved the last problem by using letters to distinguish the several forms when he obtained them under one name, since it was impossible to determine which of them was properly entitled to the name. This confusion was unfortunate, because the several forms appearing under one name often included both worthless types and types of considerable merit, as for example with the six "Red Antwerp" types, which Grubb designated "Red Antwerp A" to "Red Antwerp F". An example of a cultivar with several synonymous names was "Bath's Perfection"; Grubb found it identical with a stock received as "Laxton's Abundance" but it was probably the old American cultivar "Marlboro".

Grubb considered that the stock that he designated "Baumforth's A" was probably the original "Baumforth" cultivar of 1865, and that the one he designated "Baumforth's B" was most likely the original "Northumberland Fillbasket", though Card (1898) stated that it was a seedling of this cultivar. It was distinct from the so-called "Kirriemuir Fillbasket" and "Laxton's Fillbasket". Grubb noted that "Bath's Perfection" ("Marlboro") was the cultivar most commonly grown in England, though it did not do well in Scotland, and was prone to a "die-back" disease. "Baumforth's B" was widely grown in Scotland, where it had the reputation of being widely adapted and more successful than most cultivars on poor soil; it was, however, severely affected by cane spot or "Baumforth's disease". The other popular cultivar in Scotland was "Mitchell's Seedling", which Grubb did not find growing elsewhere, though he was not certain whether its origin was Scottish. It was sometimes called "Semper Fidelis", but it is unlikely that it was the same "Semper Fidelis" that Carters introduced in 1885.

It is interesting to note the relative performance of some 10 cultivars which Grubb tested from 1919 to 1922. These are given in Table 2.2. No doubt some of the stocks were degenerate through virus infection, and Grubb's results do not represent their potential when first released.

In a later paper Grubb ranked seven different cultivars in the following order: "Baumforth's A", "Helston", "Lloyd George", "Red Antwerp C", "North Ward", "Red Cross", "Pyne's Royal" (Grubb, 1931).

Table 2.2. The performance of raspberry cultivars at East Malling Research Station from 1919 to 1922 relative to that of "Pyne's Royal" (Grubb, 1922)

Pyne's Royal	100
Baumforth's B	85
Bunyard's Profusion	73
Devon	69
Norwich Wonder	64
Bath's Perfection	60
Baumforth's C	58
Superlative	42
Hornet B	39
Park Lane	33

2.4 EARLY RASPBERRY CULTIVARS IN NORTH AMERICA

In North America the red raspberries cultivated before 1800 were of European origin. M'Mahon recommended the English raspberry in the American Gardeners' Calendar for 1806, and of the 20 cultivars mentioned in Prince's *Pomological Manual* of 1832, 16 were probably from *R. idaeus*, only three from *R. strigosus* and one was a hybrid of the two types. In 1853, the American Pomological Society recommended four *R. idaeus* cultivars for general cultivation; in 1891 it recommended 14 from *R. idaeus* and 6 from *R. strigosus*. It was generally thought that *R. idaeus* forms were superior in fruit quality, but they proved less adapted than the local forms to the extremes of summer and winter weather encountered in America. The most popular of the European forms was "Red Antwerp", which was praised by M'Mahon in 1806, and, though it did not become popular until later, it was important enough to be reported upon in the U.S. Department of Agriculture's report for 1866. "Brinckle's Orange" was raised by W. D. Brinckle of Philadelphia in 1845, and was regarded as the best kind for flavour and quality until the early 1900s. It was derived from a large-fruited English cultivar called "Dyack's Seedling", but the pollen source is not known: most probably it was pollen of a native *R. strigosus* form.

Although *R. idaeus* forms were preferred, wild forms of *R. strigosus* were harvested and some were selected for cultivation. Among the first were "Stoever", found in Vermont about 1859, "Brandywine" (of unknown origin) and "Turner", raised in about 1850 in Illinois. However, "Marlboro", produced by A. J. Caywood of Marlborough, New York was the most famous and quickly became renowned for its large fruit and good hardiness. A selection from Mount Mitchell in North Carolina was found to be resistant to the root rots found in Oregon and was used extensively in breeding there.

The use of controlled crosses for breeding new cultivars became important at an earlier date in America than in Europe. An early example is "Latham", which was selected from a cross between "King" and "Loudon" and introduced commercially by the Minnesota Fruit Breeding Farm in 1914. "Latham" remained the leading cultivar in eastern America until the introduction in 1930 of "Newburgh" from Geneva Experimental Station, New York, and its own derivative "Chief", also selected at Minnesota. "Latham" and "Chief" later became popular in parts of eastern Europe where hardiness is an important requirement.

2.5 SOME HYBRIDS BETWEEN NORTH AMERICAN AND EUROPEAN RED RASPBERRIES

Great advances in raspberry improvement occurred when European cultivars of the *Rubus idaeus* type were crossed with American cultivars of the *R. strigosus* type. "Brinckle's Orange" may well have been one of these, but the outstanding example was "Cuthbert", which was discovered about 1865 at

Riverside, New York by Thomas Cuthbert and was almost certainly derived from a cross between "Hudson River Antwerp" (*R. idaeus*) and a native *R. strigosus* type. It was a leading cultivar from the 1890s to the 1940s, particularly on the west coast of America. Indeed, it was not seriously challenged until "Washington", one of its progeny from a cross with "Lloyd George", was introduced from Puyallup in 1943. Another success from a "Cuthbert" × "Lloyd George" cross was "Early Red", released from South Haven, Michigan in 1952, and a successful selection from a cross of "Cuthbert" with "Willamette" was "Meeker" introduced in 1967 from Puyallup. "Washington" itself was the parent of the three successful cultivars: "Puyallup", "Sumner" and "Fairview" (see Appendix 1).

In Europe a similar success was achieved by "Preussen" (synonym "Berlin"), which was introduced into Germany in 1919 and thought to be derived from a cross between "Superlative" (*R. idaeus*) and "Marlboro" (*R. strigosus*). Preussen itself was not very successful as a cultivar, possibly because it tends to produce too few canes, but it has been remarkably successful as a parent, particularly in combination with "Lloyd George".

Crossing between European and American forms became even more successful when "Lloyd George" was widely used as a parent. This cultivar contributed new germplasm for primocane fruiting, large fruit size, and resistance to the American aphid vector of virus disease. The improvement in fruit size of some of its hybrids was possibly achieved because they combined the long-conical shape of the "Lloyd George" receptacle with the more rounded shape of American raspberries. The outstanding example is "Willamette", which George Waldo selected from the cross "Newburgh" × "Lloyd George" and released in 1943. "Willamette" soon dominated raspberry production in western America and has done so until the present day.

The impact on raspberry improvement of "Lloyd George" was emphasized by Oydvin (1970), who observed that 32% of North American and European cultivars had "Lloyd George" as one parent, 37% were less closely related and only 31% were unrelated. In America it was used successfully in breeding at Geneva (New York), Corvallis (Oregon), Puyallup (Washington), Knoxville (Tennessee), South Haven (Michigan) and Ottawa. A particularly successful combination was the cross of "Lloyd George" with "Newman 23", from which at least seven cultivars were selected at Ottawa and Geneva; "Newman 23" itself is a *R. strigosus* type introduced about 1924 by C. P. Newman of Quebec, possibly from a cross between "Eaton" and "King": it was considered outstanding for its fruit firmness and hardiness. In Europe, crossing "Lloyd George" with "Preussen" gave the cultivars "Malling Notable", "Malling M", "Norna", "Rubin", "Schonemann" and "Paul Camarzind" (Appendix 1). "Lloyd

George" also combined well with "Pyne's Royal" and breeding at East Malling Research Station with "Lloyd George" × "Pyne's Royal" derivatives eventually produced the cultivars "Malling Jewel", "Malling Exploit" and "Malling Promise". In Holland the cross between "Lloyd George" and "Hornet", a French cultivar dating from about 1850, gave "St. Walfreid", a large-fruited cultivar introduced in 1931.

2.6 EUROPEAN RASPBERRY CULTIVARS INTRODUCED FROM 1945 TO 1970

In the early days of raspberry domestication, new cultivars usually made their impact because they showed big improvements in fruit size or yield. Later the need was for vigorous cultivars to replace those that had become degenerate through virus infection, a tendency which was accelerated by the 1939–45 war, when little attention was paid to the maintenance of healthy stocks of luxury crops like the raspberry. There was also a demand for cultivars to meet particular requirements, such as suitability for processing or for travel to distant markets, for which greater firmness of fruit texture was required; early- and late-fruiting cultivars were also needed to complement the main-season types at either end of the season, late cultivars being of particular interest to strawberry growers who prefer to pick the peak of their strawberry crop before turning to raspberries.

"Lloyd George" met many of these needs, because its fruits ripen from early to mid-season and are of high enough quality for the processors, but it tended to become rapidly infected by viruses and few of the stocks available in the late 1930s and 1940s were productive. Fortunately it had been exported to New Zealand, where no major aphid vectors of virus diseases occur, and so it was possible to find a clean stock for re-introduction to Britain in 1945 (Cadman, 1948; Hudson, 1947). "Norfolk Giant" was another cultivar that became popular in Britain, both before and after the war. It had good fruit quality and its late ripening suited the cropping programme of fruit growers in southern England. Its aphid resistance, conferred by minor rather than by major genes (Knight *et al.*, 1959; Jennings, 1963b), apparently protected it from rapid virus degeneration during the war years. Growers in the west of Scotland favoured "Burnetholm", a cultivar of unknown parentage originating from Lanarkshire in Scotland, because its small, very firm fruits were ideal for sending to distant markets. This cultivar is closely related to "Colin's Defiant" which is grown in Tasmania.

Few, if any, healthy stocks of other cultivars could be obtained in Europe in the immediate post-war period. It was therefore fortunate that Grubb had produced a series of advanced selections at East Malling Research Station from crosses made in the 1930s (Grubb, 1935). These were tested immediately after the war, and several were later named; one of them, "Malling Promise", from a cross between "Newburgh" and a "Lloyd George" × "Pyne's Royal" derivative, was sufficiently promising to be introduced into commerce with minimal testing in 1946. It produced larger fruits than any form known at the time, was very high yielding and filled the need for an early-ripening cultivar. "Malling Exploit", from the same cross, was released later, but it was not so successful in Britain, owing partly to its more spreading habit of growth and partly to its more uneven fruit shape. It grew well in dry areas and is adapted to southern parts of Europe, where it has become a leading cultivar. The most successful of Grubb's cultivars in Britain was "Malling Jewel", introduced in 1950. Although it was not immediately popular, it later increased in popularity and dominated British production from the mid-1950s to the early 1980s. Its considerable merits include its erect and sparsely spined canes of adequate though moderate vigour, which are easy to manage and do not conceal the fruit or impede the pickers at harvest time; its firm good-sized fruit, which are suited to all market outlets, and its tolerance of infection by several important viruses, which enables it to maintain a good performance over many years.

2.7 RASPBERRY CULTIVARS INTRODUCED AFTER 1970

As soon as the British industry became reorganized in the post-war years, the health of all important raspberry stocks was safeguarded by a certification scheme, following heat therapy to free some of them of virus infection. This meant that breeders were no longer obliged to release new cultivars merely to replace older ones which had become degenerate: new cultivars would be acceptable only if they were of superior merit to the existing ones. "Glen Clova" (Fig. 2.2), bred at the Scottish Horticultural Research Institute and released in 1969, became established as the predominant early cultivar, largely replacing "Malling Promise" and then "Malling Jewel". Its advantage over "Malling Promise" was its combination of high yield with better fruit quality, which meant that it could be used for canning and freezing, for which "Malling Promise" is not so well suited. Its fruit ripen over a long season and it usually outyields "Malling Jewel", especially in the early years after planting. Hence it soon became the dominant mid-season cultivar as well. Its limitation was vigorous vegetative growth which made picking and other operations difficult. This problem was eventually solved by the introduction of a cultural method of cane vigour control. The standard set by "Malling Jewel" for good growth habit was not equalled by

Fig. 2.2. Fruit of "Glen Clova" raspberry: a cultivar bred at the Scottish Horticultural Research Institute (now Scottish Crop Research Institute) and the most widely grown cultivar in Britain in 1987.

any of the new European cultivars. The two East Malling cultivars "Malling Orion" and "Malling Admiral", introduced in 1970 and 1971 respectively, also compare unfavourably with "Malling Jewel" for their growth habit. "Malling Admiral" proved a useful late cultivar and largely replaced "Norfolk Giant" because of its superior yield and fruit quality. "Leo", released by East Malling in 1975, fulfilled a special requirement, because it is distinctly later than any other cultivar and provided a significant extension of the growing season. "Joy", released in 1980, is nearly as late but does not have such good quality as "Leo". "Delight", released from East Malling in 1974, has long and conical fruit which are bigger than any other cultivar; they are soft and pale and not well suited for processing, though the large size is an advantage for the dessert trade.

"Glen Moy" and "Glen Prosen", the first spine-free raspberries, were released by the Scottish Crop Research Institute in 1981 (Fig. 2.3). "Glen Moy" is an early cultivar and provides an alternative to "Glen Clova" with notable improvements in fruit size and flavour, and "Glen Prosen" is a late cultivar with notable improvements in the firmness of fruit texture.

Fig. 2.3. Fruit of "Glen Prosen" raspberry; a cultivar from the Scottish Crop Research Institute noted for its very firm fruit, an improvement attributed to genes introduced from the black raspberry.

Considerable breeding was also done in Germany, particularly at the Max-Planck Institute. The late-ripening "Schonemann" from the cross "Lloyd George" × "Preussen" was the most notable success and became widely grown. Other successes bred at this Institute include the aphid-resistant cultivars "Rucami", "Rumilo", "Rusilva", "Ruku" and "Rutrago" (Bauer, 1980 and unpublished work).

In eastern Europe, notably in Poland, Hungary, Yugoslavia and Bulgaria, a large increase in raspberry acreage occurred in the 1950s, 1960s and 1970s. Many cultivars were tried and "Malling Exploit" and "Malling Promise" were among the most successful, partly because of their earliness, large fruit size and good yield potential, and partly because "Malling Exploit" proved tolerant of the hot and dry summers encountered. "Krupna Dvoroda" and "Gradinia" were bred at the Fruit Research Institute at Čačak in Yugoslavia from crossing "Malling Exploit" with "Rubin". They were introduced in 1973 and showed the good features of "Malling Exploit" with improvements in fruit size and quality.

In Bulgaria "Rubin Bulgarski", sometimes referred to as "Bulgarian Rubin", was widely grown, together with "Newburgh". Breeding work at the Small Fruits Station at Kostinbrod later produced "Shopska Alena", which was popular for fresh market sales, "Ralica", which has black raspberry germplasm in its ancestry, "Iskia" and the primocane-fruiting cultivar "Ljulin". "Malling Exploit" was also widely grown in Hungary, but the small-fruited Hungarian cultivar "Nagymarosi" remained popular because of its excellent flavour. Crosses between "Malling Exploit" and "Nagymarosi" were made at the Institute for Fruit and Ornamental Plant Growing, Fertöd and gave the cultivar known as "Fertöd 4".

In North America, the large production areas of Oregon, Washington and British Columbia were not affected by war-time conditions as badly as their European counterparts. The main cultivars grown in the 1950s were "Willamette" and "Canby", bred by G. F. Waldo at Corvallis, Oregon, and "Washington", "Puyallup" and "Sumner" bred by C. D. Schwartze at Puyallup, Washington. Of these "Willamette" became the most important, probably because of its superior fruit qualities, and held a central position until the 1980s, similar to that held in Britain by "Malling Jewel". "Washington" was popular at first, particularly for its superior flavour, while certain stocks of "Sumner" gave problems because of a crumbly fruit condition (see p. 123). Although all of these cultivars have "Lloyd George" in their ancestry, only "Canby" has inherited its major gene for resistance to the American aphid. "Fairview", introduced from Oregon in 1961 was expected to produce higher yields and to replace the earlier cultivars, but its excessively vigorous growth proved unmanageable and the cultivar lost favour with growers, much like some of the new cultivars released in Britain. "Meeker", introduced from Puyallup in 1967, proved more successful and replaced "Willamette" to some extent, especially for markets that require less-dark fruit.

The problem of excessively vigorous primocanes is particularly apparent in British Columbia, where the growth made by all cultivars is especially vigorous. Particular attention was paid to the problem in breeding work there by selection for cultivars of only moderate vigour. Two examples are "Matsqui" and "Haida", both of which are notable for their short, stocky canes with short internodes. Their compact growth habit makes them well suited to British Columbia, but in other areas their growth is sometimes inadequate to ensure satisfactory yields. In British Columbia the newer varieties "Skeena", "Chilcotin" and "Nootka" became more widely grown because of their high yield potential and good fruit quality. "Skeena" is another example of a successful cultivar produced by crossing between a North American and a European parent, the latter in this instance being a close relative of "Glen Clova". It has become popular because of its high yield and excellent fruiting laterals which display the fruit well and make picking easy. "Chilcotin" and "Chilliwack" are notable for the excellent bright colour of their fruit which give them exceptionally good eye-appeal and make them ideal for fresh sales.

2.8 RESTRUCTURING THE RASPBERRY FOR HIGHER YIELD

Many attempts are being made at restructuring raspberries to improve their yield potential without changing their height. Jennings (1966a) discovered a large-fruited mutant of "Malling Jewel" in a commercial field; this turned out to be caused by mutation of a major gene, designated L_1, which had pleiotropic effects on nearly all aspects of plant development. The most striking effect was to increase the sizes of the fruiting lateral and all the flower parts, especially the receptacle, whose growth was prolonged to produce an elongated cone bearing some 50% more drupelets; since each drupelet was also enlarged, the total increase in fruit size was considerable. The large fruit size of SCRI 7518E6 (Fig. 2.4) is partly attributed to gene L_1, but the gene has a tendency to mutate back to a normal form, and the resultant instability has prevented breeders from making more use of this gene. Increases in fruit size have been achieved more reliably by selecting for fruit containing more drupelets ("Delight" and "Glen Moy" are notable examples) or for a combination of large drupelets and high drupelet number as in "Krupna Dvoroda". This has often given fruit having a more long-conical shape.

Keep and Knight (1968) attempted to improve yield by interspecific crossing to transfer genes for higher numbers of fruits per fruiting lateral, and succeeded in raising the number of fruits present from about 10 or 12 to

Fig. 2.4. Fruit of SCRI 7518E6 raspberry. The large fruit of this cultivar are attributed to gene L_1, whose presence is recognized by the large sepals and stipules.

about 35 (see p. 165). The cultivar "Joy" illustrates the progress that was made by selecting for high fruit number within the germplasm of *R. idaeus*. However, this characteristic is usually a feature of late-ripening cultivars which have long fruiting laterals, and is not easily introduced into early cultivars. The number of fruiting laterals present on a cane also received attention, both in respect of the number of nodes present in the cropping zone (see p. 153) and the number of fruiting laterals present at each node (see p. 162).

2.9 PRIMOCANE FRUITING AND ITS OCCURRENCE IN *RUBUS* SPECIES

In most domesticated forms of red raspberries the axillary buds of the first year's canes remain dormant until the second year and then produce lateral fruiting branches. In primocane-fruiting forms, fruit are produced at the tips of one-year-old growth at the end of the first season, leaving only the lower part of the cane to fruit in the second season. This behaviour is exceptional in raspberries and blackberries, but it is characteristic of such *Rubus* species as *R. arcticus*, and *R. saxatilis*. Focke used the character to separate some of the herbaceous subgenera, though it is not always characteristic of particular subgenera because it occurs in individual species of some, for example in *R. illecebrosus* Focke of the *Idaeobatus* and in *R. odoratus*, L. of the *Anoplobatus*. Haskell (1960) found segregation for the character in 5 out of 20 progenies of wild raspberries raised from seed collected in Britain. Some of the wild raspberries found in Siberia, both of *Rubus idaeus* and *R. melanolasius*, are also primocane-fruiting, perhaps because primocane-fruiting is an adaptive feature enabling the canes to fruit in the severe Siberian climate (Kuminov, 1956).

Lewis (1941) noticed that many triploid and tetraploid cultivars were primocane fruiting, but he showed that this was because selection had favoured polyploids which were primocane fruiting and not because primocane fruiting was a direct expression of polyploidy. This may be because fruit development for the autumn crop occurs when the prevailing temperatures are higher and more favourable for good fruit set in subfertile polyploids. Possibly the temperature difference was more effective in France than in more northerly countries, because many tetraploid primocane-fruiting cultivars originated there in the 19th century. Some notable examples are "Fontenay aux Roses", "Belle de Fontenay" (synonym "Belle d'Orleans"), "Perpetuelle de Billiard", "Merveille de Quatre Saisons" (synonym "Merveille Rouge"), "Souvenir de Desire", "Brunean", "Surpasse", "Merveille Blanc", "Colossus" and "Surprise d'Automne". Their English counterparts are "Hailsham" and "November Abundance". "La France" originated as a seedling of a French cultivar in Connecticut, U.S.A. None of these sources of primocane fruiting has proved useful for breeding.

For many years the most important cultivar grown for autumn raspberries was "Lloyd George", which produces a good second crop in a favourable season. In Britain, Williams (1959c, 1960) investigated the initiation of flower initials in this cultivar and found that the shoots in which initiation occurred earliest were still elongating, in contrast to a non-primocane-fruiting type where the first flower initials occurred later, after shoot elongation had ceased. In "Lloyd George", Williams found that flower-bud initiation had occurred in the terminals after the third week of August. In Oregon, Waldo (1934) found that flower-bud-initiation began as soon as shoot elongation ceased in mid to late July. However, both in Britain and in America, breeders of primocane-fruiting raspberries have now selected types which initiate their flowers in June, flower in late June and July and ripen their fruit from late July until frost. The process of flower-bud initiation occurs basipetally; in some cultivars it continues rapidly down the full length of the cane, while in others it continues for 16–18 nodes or less in the autumn, leaving the remainder to initiate in the spring.

Both Haskell and Lewis (see Keep, 1961) regarded primocane fruiting as a discrete character which could be considered characteristic of all or nearly all of a plant's canes, and postulated that it was controlled by a major gene. This interpretation is not generally supported. Keep (1961) found that autumn fruiting was a continuously varying character determined quantitatively by genes acting in an additive or complementary way. She obtained proportions of primocane-fruiting progeny related to the primocane-fruiting potential of the parents used, which was also correlated with the earliness of the autumn fruits in the progenies. Moreover, the primocane-fruiting character varied in expression within the plant, some plants fruiting on only one of many canes, and it was influenced by several factors, especially plant age. She suggested that primocane fruiting occurred when flower initiation took place in the terminal bud of the cane before cane elongation ceased, and that variation in its expression was caused by interactions of each of these processes with the environment. In effect, the primocane-fruiting genotypes are day-length and temperature neutral, since they initiate their flowers in long days and high temperatures, in contrast to the short-day and lower-temperature requirements of non-autumn-fruiting genotypes. The only factor limiting their primocane fruiting seems to be a growth factor: Williams (1960), for example, found that "Lloyd George" initiated flowers in long days and high temperatures but only after seven months of growth. More recently, Lockshin and Elfving (1981) showed that for "Heritage" a combination of high temperature and high nitrogen promoted the greatest cane elongation, the earliest and most profuse flowering, and the most flowers per unit of cane growth (see also p. 151). However, primocane-fruiting and non-primocane-fruiting genotypes should not be classified into two discrete groups, because the former merely represent one end of a continuous range of day-length/temperature responses.

2.10 BREEDING PRIMOCANE-FRUITING RASPBERRIES

Many breeders have improved the earliness of primocane-fruiting raspberries. In Europe, Keep (Anon., 1969; Knight and Keep, 1966; Keep *et al.*, 1980) combined genes for autumn fruiting derived from different cultivars and selections of wild red raspberries with others derived from such diverse species as *R. odoratus*, *R. illecebrosus*, the strawberry raspberry, and *R. arcticus*. In America, Slate and Watson (1964) used "Lloyd George" extensively in breeding for large fruit size and consequently produced several cultivars with genes for primocane fruiting, notably "Indian Summer", "Taylor" and "Marcy", whose autumn fruits are rather late. However, progress at New York was slow at first, because the extensive use of "Lloyd George" and its derivatives led to inbreeding depression and because there seemed to be a genetic linkage between small fruit size and genes that affect primocane fruiting (Ourecky, 1978). Another source in North America was "Ranere", a small-fruited cultivar sometimes known as "St. Regis", which originated in New Jersey and gives an early autumn crop. Genes for primocane fruiting derived from "Ranere" or its derivative "Sunrise" combined well with those derived from "Lloyd George" and produced much earlier primocane-fruiting selections. Crosses to achieve this combination were made at Virginia (Oberle *et al.*, 1949) as well as at Geneva, New York, where the cross between "Ranere" and "Marcy" gave "September", which was the standard primocane-fruiting raspberry until the 1970s.

New advances became possible with the discovery in New York State of wild raspberries whose autumn fruits ripened in early August and whose canes branched near the top to give an increased fruiting surface. These and the cultivar "Durham", a "Taylor" derivative released in 1947 from Durham, New Hampshire, proved good sources of both early primocane fruiting and an extensively branched habit. Their progenies began ripening in early August at Geneva and had a fruit-bearing surface extending to some two-thirds of the current season's cane. This was increased further by extensive branching, with a result that their fruit production approached one-half that of summer-fruiting cultivars (Ourecky, 1975). The most successful selection was "Heritage", which was selected from a progeny of

"Durham" crossed with a "Milton" × "Cuthbert" hybrid and introduced in 1969. This cultivar is notable for its sturdy, erect canes whose laterals develop basipetally for a third or more of the canes' length. It can be grown without support in the north-eastern states of North America, but not in western or southern areas where it forms a more open habit. Its high yield of autumn fruit continues to ripen until the onset of frost, and it has remarkable fruit qualities which have made it a major success in many parts of the world.

Breeding also proceeded at Durham, New Hampshire, where "Fallred", selected from a progeny of a New Hampshire and a Geneva selection, was introduced in 1964. At Corvallis, Oregon, Waldo and Darrow (1941) found that "Lloyd George" × "Ranere" crosses gave progenies with the earliest and the highest proportions of primocane-fruiting segregates. They also crossed "Lloyd George" with "Cuthbert", "Newburgh", "Viking", "Chief", "Latham" and "Ranere" and obtained their best selections from the cross with "Cuthbert".

Breeding to combine a self-supporting branching habit with early autumn fruiting was emphasized in later work at Corvallis. In this work Lawrence (1976, 1980) found that two complementary genes controlled branching and that there was a correlation between early autumn flowering and branching habit, though selection for early autumn-flowering alone was not sufficient to obtain branching types. The parents used as sources of autumn fruiting were "Heritage", "Fallred", "Zeva Herbsternte", selections of *R. strigosus* origin from Wyoming, and selections from the East Malling work in Britain. Some of the Wyoming selections ripened fruit in early July, completed cropping in August and were remarkable for their self-supporting habit. This was considerable progress, but the amount of growth produced by July was not always adequate to carry a respectable yield. Two of the best self-supporting Wyoming selections were later named "Pathfinder" and "Trailblazer". In the Corvallis programme, early primocane-fruiting types were crossed with large-fruited summer-cropping cultivars to improve fruit size. These crosses gave predominantly late primocane-fruiting progenies, but subsequent intercrossing and then backcrossing to early primocane-fruiting selections gave the desired combination of characters. Selections with vigorous, self-supporting canes and firm fruits weighing 2.8 to 3.1 g were obtained and one of them was named "Amity" (Lawrence, 1980). It seems likely that future primocane-fruiting cultivars will include some whose canes are self-supporting and some whose canes need a measure of support during the growing season. Where earliness is a major objective they will also have high cane numbers and large fruit size to compensate for the paucity of lateral numbers (Hoover *et al.*, 1986).

In Europe, one of the most successful primocane-fruiting cultivars for some time was "Zeva Herbsternte". This can be regarded as a "Romy"

5 Hybrid Berries and Arctic Berries

Hybrids between distantly related *Rubus* species have played an important part in the evolution of cultivated forms. The Loganberry is easily the most important of them, and is an outstanding example of how a distinctly new form of fruit can arise by direct hybridization between two distantly related species. Not only did it become important in its own right but it also played a vital role in the domestication of the Western American blackberries, becoming the most important route by which genes determining the development of bisexual flowers were transferred to this important blackberry group: the essential first step in their improvement. The Nessberry is a similar example, but it is interesting that in this instance several generations of breeding were necessary after the original cross before a fertile cultivar was obtained. The use of hybrids to transfer specific characters from "donor" species is a normal breeding procedure for improving both raspberries and blackberries, as will be evident from the preceding chapters, but some of the hybrid berries have become so important in their own right that they must be considered separately as new kinds of fruit.

5.1 LOGANBERRY

The Loganberry is the most successful hybrid berry. It was discovered in 1883 in the garden of Judge Logan at Santa Cruz, California, growing in a bed which the judge had sown with seed collected from the "Aughinbaugh" blackberry, a pistallate form of the Western American blackberry, *R. ursinus*. The judge had sown the seed in 1881 and planted out about 100 seedlings. One of these was readily distinguished at fruiting time by its reddish-purple fruit colour, though apart from having slightly larger leaves and canes and more vigorous growth it was generally similar to the others and to the "Aughinbaugh" blackberry (Fig. 5.1). Judge Logan had no doubt that the pollen parent of this seedling was a raspberry. Raspberries resembling "Red Antwerp", a cultivar of European raspberry, also grew in his garden and he was confident that his new berry was a raspberry–blackberry hybrid. It soon became very popular among his friends and in 1902 he wrote "this red fruit is universally known here as the Loganberry".

Judge Logan's account of the hybrid origin of the Loganberry was

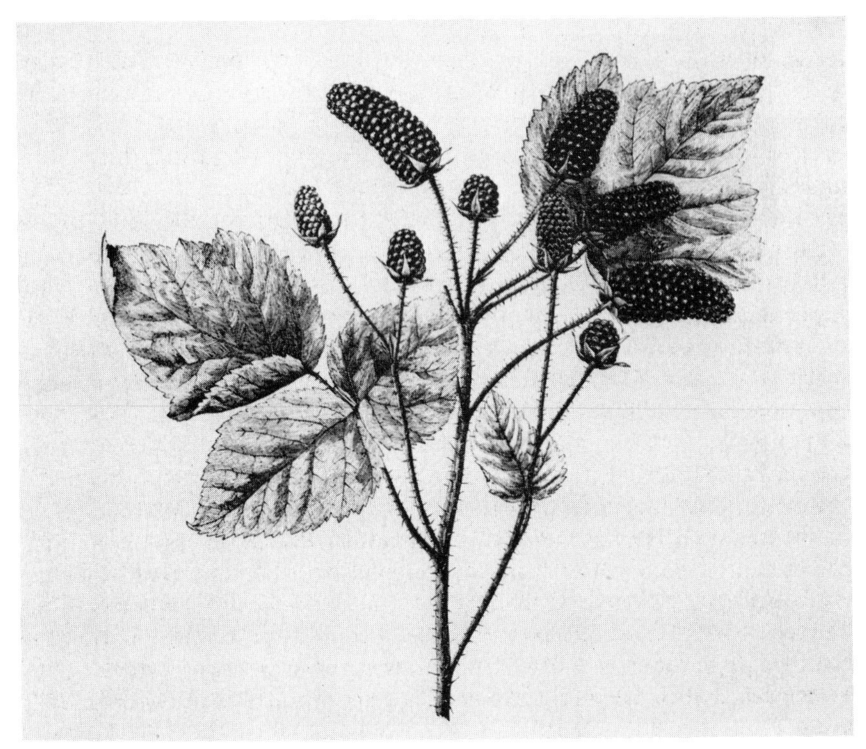

Fig. 5.1. The "Aughinbaugh" blackberry, a pistallate form of the western American blackberry, *Rubus ursinus*, which Judge Logan used as a parent of his "Loganberry" and "Mammoth" blackberry in 1881. (Reproduced from the *California Horticulturist and Floral Magazine*, Vol. 5, 1875.)

accepted at first, but then became the subject of debate. The problem was that many of its unique characters were found to be dominant in crosses with either raspberries or blackberries, suggesting the behaviour of a species. In 1906 J. H. Wilson supported this view in a paper given to an international conference in London. He reported that no truly raspberry characters had segregated in his crosses of the Loganberry with the raspberry, only intermediate characters like those of the Loganberry itself. Other workers, notably Hedrick (1925), Darrow and Longley (1933) and Darrow (1937) disputed the hybrid hypothesis because the Loganberry was highly fertile, and also because they thought that its progenies showed insufficient diversity for a hybrid obtained by a distant cross. Bailey was among those who favoured the hybrid origin, though he gave the Loganberry the specific rank of *R. loganobaccus* in his systemic studies. The subject was debated for many years, but in 1940 cytological evidence from the John Innes Horticultural

Institute in England both supported the hybrid origin of the Loganberry and explained its true-breeding behaviour (Crane, 1940; Thomas, 1940a). Crane and Thomas showed that the Loganberry was a hexaploid with 42 somatic chromosomes which associated in 21 bivalent pairs at meiosis. They made crosses among the Loganberry, the raspberry and *R. vitifolius* (a form of *R. ursinus*), and from cytological studies of the resultant hybrids concluded that the Loganberry was an allopolyploid of constitution $v_1v_1v_2v_2\ ii$; v_1 and v_2 denoting two differentiated genomes derived from *R. vitifolius* and i a genome derived from *R. idaeus*. This meant that it was produced because an unreduced pollen grain of the red raspberry had functioned in just the kind of cross that Judge Logan had speculated, and his explanation therefore became generally accepted.

Crane and Thomas were also among the first to resynthesize the Loganberry. From the cross of the tetraploid red raspberry "Hailsham" with the octoploid western American blackberry *R. vitifolius* they obtained a progeny which was uniformly hexaploid and fertile, and resembled the Loganberry in essential botanical characters, though it did not contain individuals of horticultural merit. A similar attempt was made in the U.S.A. by Waldo and Darrow (1948), who crossed the tetraploid raspberry "La France" with each of nine forms of trailing western blackberry. Similarity with the Loganberry was again a feature of the progenies obtained, and selections with merit were obtained in this work, one in particular having fruits with an average weight of 6.2 g. Other workers, including the present author, have had similar successes (see p. 64), which all support Judge Logan's explanation of the origin of his Loganberry. They also show what amazing luck the Judge had. Diploid pollen grains are produced very infrequently by diploid raspberries, and yet one of them not only reached the stigma of an "Aughinbaugh" flower in the Judge's garden but produced a hybrid which remained unsurpassed for nearly a century, despite many hundreds, possibly thousands, of similar hybrids that were subsequently raised by controlled pollinations. No doubt the "Aughinbaugh" was a superior parent to the blackberries used later.

A characteristic of allopolyploids like the Loganberry is their tendency to breed within well defined limits, and both the spontaneous and resynthesized Loganberries segregate within the limits of the variation present within their parent species, and not for the characters which distinguish these species. Seedling derivatives of the Loganberry have consequently been propagated widely under the Loganberry name, and Loganberry stocks have become variable for season of ripening, vigour, fruit size and leaf shape. Because of these and other variations caused by virus infection or mutation, it became necessary to sort out the many clones available in commerce. This was done by Beakbane (1939, 1941) at the East Malling

Research Station, England in the early 1930s. She found that the stocks available showed a considerable admixture of inferior types, and that some of them were degenerate because of virus infection. Several clones apparently identical to the original Loganberry and apparently free of virus were selected for yield trial, and the best was propagated vegetatively and distributed as clone number LY59.

The only important mutation of the Loganberry is the spine-free form discovered in California by B. E. and G. R. Bauer in 1929 (Brooks and Olmo, 1944). This mutation, originally referred to as "Bauer's Thornless", is similar to the Loganberry except for being spine-free and slightly earlier. It was distributed in America in 1934 and reached England about 1937, but it was slow to become popular because of the commonly held belief that it was inferior to the spiny form. However, Way (1967) at East Malling reported that two of the spine-free clones in commerce differed from each other, and that one of them, number L654, gave a higher yield in a trial than LY59, the best of the spiny clones. It achieved this by producing more canes per plant and a larger average fruit size: otherwise it was morphologically similar to the spiny Loganberry. Like several of the spine-free blackberry mutants, however, the mutated tissue is confined to the outer layer (L1) of the meristem and therefore remained inaccessible to plant breeders until the 1980s (see p. 159).

The Loganberry achieved immediate success because of its large size, attractive conical shape and characteristic flavour. Nevertheless, its colour and softness are not ideal for freezing, and its flavour is too acid for some purposes; indeed, the fruit has to be very ripe before its acidity is low enough for eating fresh. Hence it did not meet with sustained consumer acceptance. Its tendency to produce only moderate yields have also tended to limit its use on a large scale, though this problem has become less serious in Britain now that trustworthy stocks are available. Largely because of these and other shortcomings, efforts were made to produce superior blackberry cultivars. This work is described in Chapter 4, but it is worth noting here the considerable role played by the Loganberry in conferring perfect flowers (with male and female parts) to the native western American blackberries. Indeed, it is possible that the blackberries "Santiam", "Johnson", "Starr" and "Lincoln", which were selected from the wild as perfect-flowered variants of the native dioecious species and figure prominently in early breeding work, were in fact natural hybrids between the Loganberry and native forms. They resemble plants subsequently obtained by controlled crossing of this kind.

Crosses between the Loganberry and the raspberry are not successful and usually produce entirely sterile progenies. Messrs Laxtons, a nursery firm in England, crossed the Loganberry with the raspberry "Superlative" and introduced a selection which they named "Laxtonberry" in 1906. It was the most fertile individual in a large family and had 49 chromosomes, presumably made up of an extra raspberry genome in addition to the normal complement of the Loganberry, and formed by the fusion of a raspberry pollen grain with an unreduced egg cell of the Loganberry. Its flavour was good but its fertility was so poor that it crumbled when picked and did not become popular (Crane, 1935).

5.2 PHENOMENAL BERRY

The Phenomenal berry, sometimes known as "Burbank's Logan", was selected by Luther Burbank in California from the second generation of a cross between the "Aughinbaugh" blackberry and "Cuthbert" raspberry. It was introduced in 1905 and reached England about 1910. Although it is similar to the Loganberry, its fruit are larger, slightly lighter in colour and its advocates considered it sweeter, richer and more distinctive in flavour. But it did not compete successfully with the Loganberry in commerce. In a trial at East Malling from 1936 to 1939 it was inferior to the Loganberry for yield but superior to it for fruit size and resistance to cane spot (Beakbane, 1939). There was no difference between the two forms in canning quality, though the larger size and better appearance of the Phenomenal berries made them more suitable for some purposes. Like the Loganberry, the Phenomenal berry is a hexaploid. Its chromosome behaviour at meiosis is essentially regular, usually with 21 bivalents (Thompson, 1961).

5.3 YOUNGBERRY, BOYSENBERRY AND NECTARBERRY

The Youngberry was selected by B. M. Young of Morgan City, Louisiana in about 1905 from a cross between the hexaploid Phenomenal berry and octoploid "Austin Mayes" dewberry, which is thought to be a hybrid of *R. baileyanus* with *R. argutus*. It was not introduced until 1926, but rapidly became popular for its excellent flavour and large fruit. It has a deep wine colour and is much sweeter than the Loganberry, but though it is considered superior to it for freezing, jam making and dessert fruit, it replaced it only in parts of California and rarely in Oregon and Washington, possibly because its canes are prone to winter injury in the more northern parts. It is also preferred to the Loganberry in South Africa, where it is known as the South African Loganberry.

The Youngberry is a septaploid with 49 somatic chromosomes. It is fertile even though its chromosome behaviour at meiosis is irregular. Crane (1946) found that its pollen mother cells had on average two quadrivalents, three trivalents, up to nine univalents and the remainder bivalents. Thompson (1961) noted similar irregularities and suggested the genome formula $a_1a_1a_2a_2u_1u_1i$, where a_1 and a_2 are two differentiated genomes derived from "Austin Mayes", u_1 is a *R. ursinus* genome and *i* is a *R. idaeus* genome. She also noted that the roots show mitotic instability, which possibly explains why some writers report a chromosome number of only 42. Haskell, for example, noted somatic chromosome numbers ranging from 28 to 49 (see Williams, 1957).

E. L. Pollard of California studied six thornless mutations of the Youngberry. All but one of them were worthless, having wrinkled pale foliage and too many blossoms that failed to develop fruit. The one exception was equal or possibly superior to the Youngberry, though later reports suggested that it was less productive. It also had a slightly different leaf shape (Darrow, 1929).

The Boysenberry is similar to the Youngberry. It was selected by Ralf Boysen of California about 1920 and not introduced until 1935. Its origin is unknown, but its type is as might be expected from a cross between the Loganberry and a trailing blackberry such as "Lucretia" or "Austin Mayes". It has 49 chromosomes and irregular chromosome behaviour like the Youngberry (Thompson, 1961). Its fruit are larger, have a more deep purple colour and a more acid taste. They ripen distinctly earlier and tend to be rather soft for long-distance travel, but are considered satisfactory for canning and freezing. The Boysenberry is preferred to the Youngberry, has largely replaced the Loganberry in Oregon and Washington and attained major importance in New Zealand.

Clonal variation has been observed in the Boysenberry to a greater extent than in the Youngberry or Loganberry. Both in Oregon and New Zealand variants have been identified, evaluated in trials and the best of them propagated for commercial use. The "Riwaka" Boysenberry clone of New Zealand is being exploited the most widely (Anon., 1984).

The Nectarberry was introduced in 1937 by H. G. Benedict of California. It was reputed to be a seedling of the Youngberry, but it is so similar to the Boysenberry that it is more likely to be a mutant of it.

5.4 TAYBERRY, TUMMELBERRY, SUNBERRY AND FERTÖDI BÖTERMÖ

The success of the Loganberry is undoubtedly due to the exceptional qualities of its parents, especially those of the "Aughinbaugh" blackberry, which seems to have been a superior parent to the blackberry parents later used by breeders. However, the release of "Aurora" from Oregon in 1961 (see p. 54) provided another octoploid blackberry with excellent fruit qualities, and selection for improved tetraploid raspberries in Scotland yielded other potentially superior parents. There was therefore a new opportunity for creating hybrid berries of the Logan type by crossing between "Aurora" and some of these tetraploid raspberries. The Tayberry (Fig. 5.2) was the most promising selection of such a cross, and was released by the Scottish Horticultural Research Institute in 1979. It resembles the Loganberry in plant habit and fruit quality, but it ripens its fruit earlier, is productive and has larger and less acid fruit. Moreover, it presents its crop well on short, strong laterals, making it easy to pick. These qualities have made it popular, and it is now widely grown in Britain, Europe and North America.

The Tummelberry is a close relative of the Tayberry, produced by crossing the latter with one of its sister seedlings. It is more hardy than the Tayberry and, compared to the Tayberry, its fruit are slightly smaller, more round in shape, less intensely purple and have a less aromatic flavour. Its fruit start to ripen a week later and continue for longer. The two hybrids therefore complement each other to provide a choice of flavours and an extended season of ripening.

The Sunberry was bred at East Malling Research Station by crossing a wild form of *R. ursinus* with a raspberry obtained by selfing a tetraploid mutant of "Malling Jewel" (Keep *et al.*, 1982). The fruit are similar in size to

Fig. 5.2. A cluster of Tayberries, a hybrid fruit from a cross between the octoploid blackberry "Aurora" and an unnamed selection of tetraploid raspberry.

the Loganberry, but they are less conical in shape and more deeply coloured, though bright.

Fertödi Bötermö is a hexaploid blackberry–raspberry hybrid bred in Hungary. A chromosome-doubled (hexaploid) hybrid of *R. caesius* and "Lloyd George" raspberry was crossed with the Loganberry, and Fertödi Bötermö was selected after three generations of further breeding. It has a different, less aromatic flavour than the hybrid fruits related to the blackberries of western North America, but its fruit are large and conical, like those of the Loganberry, and the yield potential is high.

5.5 TETRAPLOID HYBRIDS OF RASPBERRY AND BLACKBERRY

The blackberry × raspberry hybrids described so far are all hexaploids or septaploids. Fewer successful hybrids have been produced at lower ploidy levels, largely because hybrids between diploid raspberries and diploid blackberries usually die as seedlings, and those between diploid raspberries and tetraploid blackberries are always subfertile. Hybrids between tetraploid raspberries and blackberries are easy to produce if the raspberry is used as female parent, but they are not useful as cultivars; their fruit tend to be a purple, almost black colour and highly pubescent; they usually separate from the receptacle in an intermediate way and they sometimes have an astringent flavour (Einsett and Pratt, 1954; D. L. Jennings, unpublished work). However, the Nessberry and Veitchberry are examples of tetraploid hybrids which proved invaluable for further breeding.

5.6 NESSBERRY

The Nessberry was derived from breeding work begun by H. Ness at the Texas Experimental Station in 1890. In 1912 he crossed a selection of the southern dewberry, *R. rubristeus* Rydb. or *R. trivialis* Michx., with the red raspberry "Brilliant" and produced 21 F_1 plants which were almost sterile, though they gave sufficient open-pollinated seed for a second generation of 280 plants. Five vigorous and fertile plants with raspberry-like characters stood out, and three of them were used to raise a large third generation, which showed little segregation. Four plants of this generation were bulked and propagated to be introduced as the Nessberry in 1921 (Ness, 1925). The hybrid had deep, blood-red fruit with a mildly acid, raspberry-like flavour. Ness hoped that it would provide for southern America what Logan's berry had provided for the Pacific Coast. But it was difficult to pick, because the drupelets adhered to the receptacle like a blackberry, and the receptacle adhered to the calyx like a raspberry. Hence the calyx was picked with the fruit as in the strawberry. Nevertheless the hybrid was successful in combining the flavour of a raspberry with the drought resistance of the southern dewberry. This was a useful advance for Texas, where most raspberry cultivars are ill-adapted and often short-lived, and so the hybrid was used extensively in breeding. Its most valuable derivative is the blackberry "Brazos", which became widely grown and has also been used extensively in breeding (see p. 52).

5.7 VEITCHBERRY AND BEDFORD GIANT

The Veitchberry was raised by Messrs Veitch of Bedford, England by crossing an English hedgerow blackberry, probably *Rubus rusticanus*, with the raspberry "November Abundance" (Crane, 1935; Crane and Darlington, 1927). This raspberry was introduced by Messrs Veitch in 1902 and was probably tetraploid at that time, though more recently it has been reported as triploid by some workers and tetraploid by others, possibly because different forms of it have arisen by contamination with self-sown seedlings or by mitotic instability. However, whatever the ploidy of the raspberry parent, a diploid pollen grain probably fused with an unreduced egg cell of the diploid blackberry to give the tetraploid seedling which was selected as the Veitchberry.

The Veitchberry has a growth habit rather like the Loganberry, but it has darker fruit and lacks the Loganberry's attractive flavour. Its biggest failing is its method of fruit separation from the receptacle, which, like the Nessberry, is intermediate between that of its two parents; hence it is difficult to pick fruit without damaging it. When selfed it breeds true except for intraspecies differences, but failure to segregate for the species differences is to be expected for an allopolyploid. However, Crane and Thomas (1949) found a much higher frequency of univalents at meiosis than expected for an allopolyploid, and suggested that this was because of imbalance among genetic factors which control chromosome pairing.

The Veitchberry was never widely grown, but it must be regarded as an important hybrid because it is a parent of the outstandingly successful blackberry, "Bedford Giant" (Crane, 1935). This cultivar was raised by Messrs Laxton of Bedford in the mid-1930s and became the most widely grown cultivar in England, partly because of its productiveness and the excellent quality of its large, round, deep-black fruit, but also because it ripens considerably earlier than any other available cultivar. It is a hexaploid, believed to be a self of the Veitchberry and formed by the fusion of a reduced and an unreduced gamete. No doubt the raspberry germplasm of its parent plays a large part in determining its early ripening.

5.8 MAHDI, MERTONBERRY AND KING'S ACRE BERRY

The Mahdi or Mahdiberry is another hybrid obtained by Messrs Veitch from crossing an English hedgerow blackberry, probably *R. rusticanus*, with a raspberry (Crane, 1935). The raspberry used in this instance was "Belle de Fontenay", another cultivar which now exists in both triploid and tetraploid forms, and since *R. rusticanus* is diploid, the "Mahdi", which is triploid, must have two genomes from one parent and one from the other. Its chromosome behaviour at meiosis is irregular and its fertility is consequently low, usually with less than 10 drupelets set per fruit. A vigorous shoot which arose from the base of a Mahdi plant was found to be pentaploid (Crane and Thomas, 1949). It set large fruit when cross-pollinated and it was named the "Mertonberry".

The Mahdi was used for breeding by Slate at New York Experimental Station, but without success (Darrow, 1937). Another triploid blackberry–raspberry hybrid from England was the King's Acre Berry, but this did not attain any importance either.

5.9 ARCTIC BERRIES

An interesting group of *Rubus* species occurs in north circumpolar or alpine regions. They are all dwarf herbaceous forms, spine-free and mostly dioecious, and they all bear flowers on annual shoots which arise from underground rhizomes. The fruit have a flat rather than a conical receptacle like the red raspberry. Apart from *R. chamaemorus* (Fig. 5.3), known variously as the bakeapple berry, cloudberry, or mountain bramble, which Focke placed on its own in the subgenus *Chamaemorus*, they all belong to the subgenus *Cylactis*. Several of them are regularly harvested from the wild, and some are popular in Scandinavia for making liqueurs.

Rubus chamaemorus L. is an octoploid and is the most common of the arctic *Rubi*. It was known to early herbalists as "Chamae Rubus", meaning the ground *Rubus*, and was later given the pre-Linnaen name of *Chamaemorus Anglica*. It is entirely circumpolar, with a sub-arctic distribution which reaches further south than the other arctic species. It is noted for its high content of ascorbic acid and is an important source of income in north and eastern Finland, especially on cultivated peat fields.

Rubus arcticus L., the arctic raspberry, is renowned for the strong aroma

Fig. 5.3. Fruit of the cloudberry, *Rubus chamaemorus*, which, like other arctic raspberries, bears fruit on short annual shoots which arise from underground rhizomes.

of its fruit. Linnaeus himself wrote that "The vinous nectar of its berries frequently recruits the spirits when almost prostrate with hunger and fatigue". The species is diploid and occurs over most of sub-arctic Eurasia, mainly between 62 and 66 degrees latitude, though in America it does not occur outside the Alaska–Yukon area. The corresponding plant there is the subspecies *R. acaulis* Michx. Although universally acclaimed for its aroma, the fruit of *R. arcticus* is very small, and its ripening is so uneven that picking is very time-consuming; Larsson (1969) found that a 500 g sample contained as many as 600 to 700 fruit.

Rubus stellatus Sm. is regarded by some authorities as a subspecies of *R. arcticus* with a more limited distribution, being found mainly in north-west Alaska and Yukon, the Aleutian Islands and Kamtchatka; it could well be a hybrid between *R. arcticus* and *R. acaulis* as it is found in the middle of their distribution centres. Its firm, bright red, glossy fruit are distinctly superior to those of *R. arcticus* in all aspects except flavour and aroma. They are subacid, with what Larsson describes as a "piquant, smokey but weak *R. arcticus* flavour", and they are much bigger (only 300 to 400 berries making a 500 g sample), much easier to pick and excellent for keeping quality.

Finally, *R. saxatilis*, the stoneberry, is a tetraploid species which also occurs all over northern Eurasia and forms natural hybrids freely with *R. arcticus*, *R. idaeus* and *R. caesius*. The species is no use for domestication because its fruit are acid and lack flavour.

Attempts to domesticate *R. arcticus* date from the early 1930s. Larsson

(1969) showed that it could be domesticated successfully in northern Sweden, and Ryynanen selected two natural strains of it in Finland which were named "Mespi" and "Mesma" and introduced in 1972 (Hiirsalmi and Säkö, 1976). These are self-sterile and must be grown together to allow cross-pollination. They have not been commercial successes. Real progress in domestication began with the successful development of species hybrids.

The goal of both Swedish and Finnish breeding work was to transfer the delicious aroma of *R. arcticus* into more easily cultivated forms with better growth habits, adequate hardiness, yield potential, disease resistance and content of ascorbic acid. In Sweden, Larsson made an extensive collection of northern *Rubus* species and produced many hybrids, many of them after doubling the chromosome number of one or both of the parent species. She also crossed them with diploid and tetraploid forms of the red raspberry, which is generally not hardy enough for polar regions. She obtained the most valuable hybrids from crosses between diploid *R. arcticus* and diploid *R. stellatus*, combining the aroma of the former with the large and attractive fruit size of the latter. The choice fragrance of the *R. stellatus* flower was also inherited by the hybrids. Both the first and second generations of these crosses were fully fertile, supporting the idea that *R. arcticus* and *R. stellatus* are ecotypes of the same species. The hybrids are now widely grown and are designated *R. arcticus* L. subsp. × *stellarcticus* G. Larsson subsp. nov. Their success provides an interesting parallel with the big advance made in red raspberry improvement when the subspecies *R. idaeus* and *R. strigosus* were first crossed together. It became the practice to plant two or more cultivars of them together to facilitate cross-pollination. Thus the paired cultivars "Anna" and "Linda" were released in Sweden in 1980, "Beata" and "Sofia" followed in 1982 and "Valentina" in 1985. Similar success was obtained in Finland (Hiirsalmi and Säkö 1980).

The octoploid nature of *R. chamaemorus* makes it an obvious choice for a study aimed at producing allopolyploids, Larsson attempted to produce hexaploid hybrids by crossing it with tetraploid forms of *R. arcticus*, *R. stellatus* and *R. idaeus*. She obtained hybrids with difficulty and those that survived the seedling stage were weak, non-hardy and of no value, though there was some theoretical interest in the similarity between a hybrid of *R. chamaemorus* and *R. idaeus* and a pentaploid plant found in Sweden and thought to be a natural hybrid between these species. Evidently natural crossing between these species can and does occur.

Larsson also produced a series of polyploid hybrids between tetraploid forms of *R. arcticus* and *R. idaeus* which were fully fertile and spine-free like *R. arcticus*, unlike the diploid and triploid hybrids of these species which were sterile. Work on diploid hybrids between *R. idaeus* and *R. arcticus* was also done at the Horticultural Research Institute at Piikkio in Finland. Early generations of these hybrids were not fertile either, but a third-generation hybrid was selected and named "Merva" though not released commercially. It is probably significant that Merva's foliage, flowers and fruit type all resemble the wild raspberry, because Vaarama found that second and third generations of this cross tended to revert to *R. idaeus* characteristics. However, "Merva" fruit make good liqueurs, because they still have something of *R. arcticus* in their taste and aroma, even though in other respects the hybrid is closer to the raspberry. The hybrid is easier to cultivate than *R. arcticus* but the fruit are too small for commerce. In 1962 a cross made between "Merva" and the raspberry "Malling Promise" gave selections with better fruit size and still with a perceptible *R. arcticus* flavour and the best of them was named "Heija" in 1975. This cultivar, referred to as a nectar raspberry, gives comparable yields to raspberry cultivars grown in Finland. A sister seedling from the same cross later proved superior, especially in hardiness and growth habit, and was named "Heisa" (Hiirsalimi and Säkö, 1976, 1980).

Rubus arcticus itself behaves peculiarly in Finland. In spite of producing abundant flowers it fruits well only in a restricted area in the middle of the country. This is apparently because its pollen does not readily achieve fertilization under the low humidities encountered in summer; a few fruit are formed in the autumn as the humidity increases. Vaarama found that fertility in *R. arcticus–idaeus* hybrids was also poor and influenced by seasonal factors, and thought that the cause might be the same as the one limiting fertility in the parent species but more strongly expressed in the hybrids. The reason is not cytological, because chromosome pairing in the F_1 hybrids is nearly perfect; in fact Vaarama (1939, 1949) found more homology between chromosomes of *R. idaeus* and *R. arcticus*, which belong respectively to subgenera *Idaeobatus* and *Cylactis*, than between *R. saxatilis* and *R. arcticus*, which both belong to subgenus *Cylactis*. Focke was clearly correct in suggesting that the subgenera *Cylactis* and *Idaeobatus* are closely related phylogenetically.

6 Cytogenetics of the Genus

6.1 THE CHROMOSOMES

The genomic number of *Rubus* is seven and species representing all ploidies from diploid to duodecaploid are found in nature. The chromosomes are rather small for morphological study. For British blackberries the range in size is from 1.5 to 2.0 μm (Heslop-Harrison, 1953), and for American blackberries, the range is from 1 to 4 μm (Einset, 1947). However, Bammi (1965a) found that chromosome preparations of *R. parvifolius* spread better than those of other *Rubi*, and he was consequently able to analyse them at pachytene. He found that they were highly differentiated and that the centromeres were easy to locate, being flanked by chromatic regions where chiasma possibly did not occur. The absolute lengths of the chromosomes and the ratios of the lengths of the long and short arms formed a good basis for identifying and classifying the seven chromosomes. More recently, Pool *et al.* (1981) identified the seven chromosomes of the red raspberry by analyses at metaphase and showed that they differed from those of *R.*

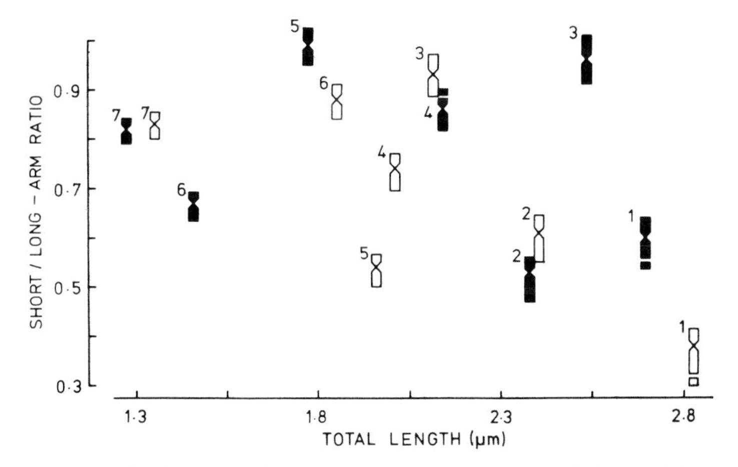

Fig. 6.1. The kariotypes of the seven chromosomes of *Rubus idaeus* (open symbols) and *Rubus coreanus* (solid symbols). Note that both species have a satellite on chromosome 1 but only *R. coreanus* has one on chromosome 4. (Reproduced from Pool *et al.* 1981.)

coreanus in many important ways, including the number of chromosomes present with satellites (Fig. 6.1). The differences were so marked that 14 different chromosomes could be recognized in F1 hybrids of the two species. The chromosomes of a range of Japanese *Rubus* species studied by Jinno (1958) also differed in the position of their centromere constrictions, and showed differences in the frequency of chromosomes with median, submedian and subterminal constrictions.

Variation in the presence or absence of chromosomes with satellites has interested many workers, but *Rubus* species show amphiplasty in the number of satellite chromosomes that they carry, and so the use of satellite number to determine species relationships and origins has not been possible. Thus Heslop-Harrison (1953) did not find more than two satellite chromosomes in all the species studied from diploids to hexaploids, and Gustafsson (1943) found satellite numbers of from one to three in species up to pentaploid.

6.2 CHROMOSOME HOMOLOGY

According to Gustafsson (1943), some 87% of the European *Moriferi* are tetraploids. Heslop-Harrison's figure of 90.7% for British taxa is close to this. Many studies of chromosome pairing have been made to determine the nature of their tetraploidy. Thomas (1940b) considered that autotetraploidy was common, but Gustafsson and others thought that alloploidy, or at least an intermediate autoalloploidy, was the rule, and that the part played by autoploidy was over-rated. The problem arises because the genomes are so similar: most of the chromosome differentiation between *Rubus* genomes appears to have occurred through minor alterations due to gene mutations, and minute structural changes where the rearrangements are so small that they do not change the ability of the chromosomes to conjugate. Hence most chromosome combinations are capable of pairing and permit crossing over, but in many hybrids the chromosome differences are sufficient to cause preferential pairing. This is shown by Thomas's work with raspberry–blackberry hybrids. He found that chromosome association in the Mahdiberry (genomes *RBB*) was almost as high as in a triploid raspberry (genomes *RRR*), but that in the Veitchberry (*RRBB*) there was regular bivalent formation due to self-pairing of the genomes. He concluded that the raspberry and blackberry genomes were sufficiently differentiated to compel self-pairing when there were two sets of each, as in the Veitchberry, but that the two genomes could pair with little restriction when there was less opportunity for preferential pairing, as in a diploid or triploid.

Heterozygosity for structural changes in the chromosomes may have complicated the interpretation of chromosome behaviour in some *Rubus* species. Translocational heterozygosity occurs in the diploid blackberries *R. tomentosus* and *R. ulmifolius*, and might well be common in tetraploids, because they frequently have only one quadrivalent at meiosis and almost exactly 50% pollen sterility: two situations more easily explained by the presence of heterozygosity for a chromosome translocation than by autoploidy (Gustafsson, 1943). "Whitford Thornless" blackberry (*R. argutus*) differs from "Burbank Thornless" blackberry (*R. ulmifolius*) by one, and possibly up to six, chromosome translocations (Hull, 1968; Ourecky, 1975), and so any tetraploid microspecies formed following crossing between these two species would indeed show meiotic irregularities. Structural diversity of the chromosomes is also indicated by chromosome inversions, and these have been recognized in such species as *R. nessensis*, *R. caesius* and *R. dissimulans* by the presence of bridges and fragments at anaphase of the first meiotic division.

Another complication is the common occurrence of secondary associations of chromosomes at meiosis, owing to their close structural relationship (Crane and Darlington, 1927; Gustafsson, 1943; Larsson, 1969). This may cause an overestimation of the frequency of quadrivalent formation and may be another cause of overestimating the occurrence of autotetraploidy.

Not withstanding these reservations it is clear that some blackberry species are essentially autotetraploids. This may be true of more American tetraploids than of European ones, because a high frequency of quadrivalents at meiosis is formed in several of them (Einset, 1947). A possible example among the European species is *R. nitidioides*, for which Thomas (1940b) found five quadrivalents at meiosis. Both triploids and pentaploids are often considered to be autoploids because of their high frequency of multivalents, but, as mentioned above, this is almost certainly because the chromosomes are not sufficiently differentiated to prevent pairing, which proceeds more freely when there are not equal sets of each. The same situation is found in triploid and pentaploid *Corylifolius* species, even though the tetraploid species behave as alloploids.

Sterility in some tetraploid blackberries is caused by a high frequency of univalents rather than by irregularity of chromosome pairing. Gustafsson attributed this to a disturbed physiological balance at meiosis. Crane and Thomas (1949) reached a similar conclusion when contrasting the relatively high chromosome pairing of "Hailsham", a tetraploid raspberry, with the Veitchberry, an allotetraploid raspberry–blackberry hybrid where failure of chromosome pairing is almost certainly due to genetically controlled physiological factors and not to lack of homology (Table 6.1).

In spite of these difficulties, some important conclusions regarding

Table 6.1. Chromosome configurations in the Veitchberry and in Hailsham raspberry (Crane and Thomas, 1949)

Chromosome configuration	Veitchberry	Hailsham
IV	0.4	3.1
III	0.1	0.6
II	11.5	5.9
I	3.1	1.6

chromosome homology can be made. Thomas (1940a) showed that the octoploid *R. vitifolius* contains four sets of each of two different genomes, designated v_1 and v_2, and that the genome of *R. idaeus* (designated i) is related to one of them. This was inferred from the rarity of univalents in a hybrid of $v_1v_1v_2v_2i$ constitution. The relationship is not close however, and sufficient preferential pairing occurs in the Loganberry ($v_1v_1v_2v_2ii$) to ensure regular meiosis. Similarly, *R. idaeus* has some affinity with *R. caesius*, which is probably allotetraploid ($c_1c_1c_2c_2$), but Rozanova (1934, 1938) produced a fertile hexaploid which was a similar cytological type to the Loganberry, having a genomic constitution of $c_1c_1c_2c_2ii$.

6.3 MITOTIC INSTABILITY AND COMPLEMENT FRACTIONATION

Mitotic instability occurs in *Rubus*, particularly among plants with a high chromosome number. Britton and Hull (1956, 1957), for example, found 28 mitotically unstable plants among 10 progenies, most of them vigorous and indistinguishable phenotypically from stable plants. The essential feature of mitosis in these plants was the grouping of the cell's chromosomes into two or more metaphase plates, each with its own spindle. Clonal propagation of the plants by root cuttings gave an array of types which were usually aneuploid and mitotically stable, though when they were unstable they tended to have a predominant chromosome number. Leaf-bud cuttings gave plants like the mother plant and perpetuated the mitotic instability. Haskell and Tun (1961) found that the chromosome numbers of a mitotically unstable plant ranged at first from 9 to 46 with a mode of 35, and then became more stable. They suggested that mitotic instability may have an evolutionary role, and that it could have been the mechanism whereby the pentaploid Mertonberry arose from the triploid Mahdiberry, apparently as a somatic mutation (Crane and Thomas, 1949).

A similar phenomenon occurs in pollen mother cells of certain high-chromosome-number hybrids (Thompson, 1962). As in somatic cells, the chromosome complement is subdivided during division into independently operating groups within the cell. Consequently, the division products contain a variable number of chromosomes, though there is a tendency for balanced genomes to be formed. Thompson called the process "complement fractionation" and pointed to its evolutionary potential: in particular to its potential for reversing the polyploidy process, because the gametes produced usually have fewer chromosomes than the parent, and fertile plants can be produced from high polyploids following fractionation of their chromosome complement.

Bammi (1965b) studied complement fractionation in a hybrid of *R. laciniatus* and *R. procerus*. He found that both of the parent species behaved as segmental allopolyploids, usually with no more than one or two quadrivalents on a single plate at meiosis, but that cells in their hybrid often had two meiotic plates which behaved as independent units. As a result, the number of sporads formed at second telophase ranged from one to over nine; only about 30% of the cells had four, compared with 97–99% in the parents. Bammi went further than Thompson and suggested that the groups of chromosomes which separated at first metaphase belonged to different genomes. There were three main reasons for his conclusion: the number of bivalents in each metaphase group was usually seven (hence the subdivision of chromosomes could not have been at random), there were rarely more than two subgroups and the divisions of each of them had a separate spindle and lacked synchronization.

There are probably many genetic factors which affect complement fractionation. It could have breeding and evolutionary potential, because separation of the genomes of a polyploid into functional gametes could give segregants bearing an affinity to the ancestral parental species, and the genome of an ancient diploid could be released from its confinement within the constitution of a polyploid.

6.4 INCOMPATIBILITY

Self-incompatibility systems occur in some 39 Rosaceous species and it is common among many of the diploid species of *Rubus*. In any one angiosperm family the incompatibility system tends to be constant, and in the Rosaceae the system found is a homomorphic gametophytic one with multi-allelic oppositional type *S* alleles at one locus (Keep, 1968*a*). Keep noted that self-incompatibility occurs in wild forms of *R. ideaus*, *R. strigosus*, *R. spectabilis* and *R. parvifolius* of subgenus *Idaeobatus*; *R. tomentosus*, *R. rusticanus*, *R. allegheniensis*, *R. cuneifolius*, *R. trivialis*, *R. argutus*, and

R. baileyanus of subgenus *Eubatus* and *R. odoratus* of subgenus *Anoplobatus*. Except for *R. baileyanus* these species are all diploids. The occurrence of self-incompatibility in such a wide range of species can only mean that it is very widespread among diploids of the genus. Keep herself discovered self-incompatibility in a further seven members of the *Idaeobatus*, one member of the *Eubatus* and four members of the *Anoplobatus*, and showed that the incompatibility was caused by inhibition of pollen-tube growth in the styles. Tammisola and Ryxjnänen (1970) postulated a more complicated system for *R. arcticus* and suggested that at least five self-incompatibility genes were involved.

In contrast to the diploid species, all the polyploid species are self-compatible, and so are all the domesticated forms of diploid *R. idaeus*, *R. strigosus* and *R. occidentalis*. Self-fertility in the first two of these species must have arisen by mutation, but this is probably not true of the black raspberry, *R. occidentalis*, because this species is probably self-compatible in the wild. All the black raspberry cultivars that Keep tested were self-compatible, including some that have originated comparatively recently from the wild. The reported absence of inbreeding depression when black raspberries are selfed suggests that self-compatibility has been established for a long time in the species, and supports the idea that cultivars were derived from wild plants which were themselves self-compatible.

A genotype becomes self-fertile when a change to self-fertility occurs in one allele to give heterozygosity at the incompatibility locus. It then prevents self-fertilization only by pollen carrying the unchanged allele. "Lloyd George" is known to be heterozygous ($S_{fert} S_5$) at this locus, and the heterozygosity probably causes aberrant segregation of genes linked with the incompatibility locus (Keep 1984, 1985).

Crosses between red and black raspberries are interesting because they are incompatible when the red raspberry is the female parent but fertile in the reciprocal direction. Purple raspberries—hybrids of the two forms —behave like their red raspberry parent. Zych (1965) observed that pollen in incompatible crosses did not progress beyond the upper third of the style and showed a marked thickening of the wall at the tip. This is similar to Keep's observation of the behaviour of self-incompatible pollen in *Rubus idaeus*, and can be explained as an example of unilateral interspecific incompatibility as described by Lewis and Crowe (1958). These workers showed that the change from self-incompatibility to self-fertility could occur through any of three mutations of the *S* gene. In the red raspberry only the pollen of cultivated forms has changed to a self-compatible state. This means that pollen of cultivated red raspberries is not inhibited on styles of self-incompatible forms; but since the incompatibility status of the styles is unchanged they still inhibit pollen of self-compatible forms such as *Rubus occidentalis*. This incompatibility varies slightly between cultivars and can sometimes be reduced by bud pollination (Hellman *et al.*, 1982).

The discovery that self-fertility occurs in a very low proportion of plants of wild raspberry populations (Keep 1968a), suggests that mutation to *S* alleles which permit self-fertility occurs commonly and that the character appears *de novo* in wild populations. It can also be induced by irradiation of pollen (Arasu, 1968). However, as Keep pointed out, the combination found in the wild of obligate outbreeding and vegetative reproduction provides both the advantages of inbreeding and full genetic flexibility: selection pressure gives individuals capable of colonizing an area by rapid vegetative propagation and these are heterozygotes rather than members of a stable inbreeding population. Indeed, vigorous vegetative suckers of established raspberry plants are so competitive that new seedlings have little opportunity of establishing themselves unless the environment changes or virus invasion reduces the vigour of the established population. Such a change of circumstances would reduce parental competition and favour seedling populations. It would also place a premium on the variability maintained by heterozygosity.

6.5 DOMESTICATION AND THE BREEDING SYSTEM

Selection of self-compatibility genes must have occurred early in the domestication or red raspberries, because garden plants are grown in monoculture and selection of self-compatible forms would inevitably accompany the selection of plants capable of giving well-set fruits in these new circumstances. Self-compatible cultivars still differ in their ability to give a good set of druplets when selfed (Bauer, 1961; Daubeny, 1971; Redalen, 1977), but while the reason for this is not understood, it is probably due to failure of drupelet development and not to failure to achieve fertilization.

Control of heterozygosity is not lacking among cultivated raspberries which have lost their self-incompatibility system, because many of the plant processes involved in fruit and seed development are geared to function in plants having an optimum for heterozygosity, and deviations from this optimum are associated with malfunction and loss of fertility: a kind of genetic inertia has tended to prevent changes from the former heterozygous state which was built up over a long period of enforced cross-breeding. Pollen germination is one aspect of the fertilization process which is still under genetic control. There are considerable differences in the capacity of pollen to germinate on styles and this can be attributed in part to genotype of the styles and in part to interaction effects between pollen and styles. These differences are most important when pollen concentration is low (Jennings, 1974).

Flower parts develop during active growth processes, but after completion of development their growth is arrested until pollination and fertilization occur. Thus the pollen must not only grow successfully in the styles but pollination or fertilization must also initiate fruit development. It is possible that pollen stimulates growth directly, because dilution of the pollen adversely affects the development of fertilized ovules and the fruit which develops after pollination by sparse pollen is slow to ripen. Alternatively, it might stimulate development indirectly through effects arising from fertilization of the fusion nucleus, and malfunction could arise if the contribution of the male nucleus is not in harmony with that of the two maternal nuclei. Jennings (1974) noted that disharmony occurred when diploid pollen of tetraploids was used to pollinate either diploid or tetraploid mother plants: this either caused development to cease altogether after a few divisions of the fusion nucleus, or to proceed so rapidly that the size of the endocarp, endosperm and embryo eight days after pollination greatly exceeded that attained when haploid pollen was used. In these circumstances embryo development suffered in competition with endocarp development, in much the same way as it does when gibberellic acid is applied to developing fruits two to four days after pollination. Treatment with gibberellic acid induces the most infertile parents to form perfect fruits, but the excessive response of the endocarp causes the pyrenes to be either devoid of seed or to contain an embryo too small to germinate.

A good set of drupelets and optimal early development of the seed depends on an optimum stimulus for early development and this largely depends on interactions between the gametes of the two parents. The maternal genotype also has considerable influence, especially on the timing of embryo sac differentiation, late differentiation being associated with low drupelet set. But once development is underway the predominant influence of pollen and maternal factors diminishes and factors derived from the seed become important.

Jennings (1974) and Topham (1967) noted that drupelet abortion at late stages was common in blackberries but less so in raspberries, though both embryos and drupelets of both fruits could cease growth at any stage of development. Abortion between four and eight days after pollination was most common in tetraploid raspberries, and some of the failures were attributed to endosperm failure. Since large differences in endosperm development occurred when parents of unlike ploidy were crossed, it was suggested that infertility in parents of like ploidy was due to an unbalanced endosperm condition similar to that induced by differences in ploidy. This would be similar to the phenomenon for which Stephens (1942) used the term "genetic strength", and described a genotype as having greater genetic strength than another when it behaved as if it were of higher ploidy in crosses with it. Topham's results suggested that the strengths of male and female gametes in *Rubus* differed, and that the change from diploid to tetraploid increased the strength of the pollen more than that of the female gamete. By postulating differences in strength among four parents, she was able to explain poor embryo development in some crosses and differences between reciprocal crosses. The reasons for these differences in so-called genetic strength are not known, but clearly they concern the roles of the two parental contributions to fruit and seed development.

The concept of differences in the strength of a parent's gametes may have its most far-reaching application when applied to pseudogamous blackberries. In the blackberry *R. laciniatus*, there are reasons to believe that sexual embryo sacs have lower strength than non-sexual diploidized ones, even when they are both tetraploid. Hence the proportion of sexual to non-sexual genotypes in a progeny depends upon the pollen parent: pollen of low ploidy tends to favour the development of sexual progeny (see p. 84). In raspberries the status of the endosperm was assessed by its response to an auxin application. A large drop in this status was found after one generation of inbreeding and a complete restoration occurred after one generation of outcrossing. Hence, on this assessment, low endosperm strength was genetically recessive and the inheritance of the differences appeared not to be complex.

Jennings (1974) also described how variations in fruit and seed development induce variations in the percentage emergence of seedlings and in the timing of their emergence: such variations govern the breeding system, because the emergence of some genotypes is favoured more than others. Two factors are involved: persisting dormancy, determined by maternal effects on the endosperm, and inbreeding effects on the embryo. But inbreeding effects predominate when dormancy is not a limiting factor: small inbred embryos show reduced or delayed germination and are at a serious disadvantage relative to earlier and more freely emerging cross-bred ones. There is thus selection for heterozygotes. Moreover, the heterozygous advantage is strengthened when the effect on the embryos is accentuated by several generations of inbreeding.

Another suggestion that selection for heterozygotes occurs was made by Keep (1969a), who considered that the large number of deleterious recessive genes found in raspberry cultivars could best be explained by postulating either some advantage to the heterozygote itself or to closely linked genes. However, there is also evidence of a balanced lethal system in one chromosome pair (Jennings, 1967a), possibly maintained in part by heterozygosity for a wild-type self-incompatibility allele (Keep, 1985). Such a system strengthens heterozygous advantage, and the chromosome concerned was found to contain an accumulation of deleterious recessive genes which had clearly been sheltered from the effects of selection.

In general, it seems that a critical co-ordination of development is necessary at each stage between the several tissues of the fruit and seed. Any deviation from the normal genetic make-up of the parents disrupts the co-ordination usually present and can lead to abortion at some stage or other. The proportion of inbred progeny that survive is reduced by the selective advantages of heterozygous embryos, and hence breeding behaviour is kept within defined limits even in the absence of a self-compatibility system.

6.6 REPRODUCTIVE BEHAVIOUR IN THE *MORIFERI* SECTION

All tetraploid blackberries are self-compatible but some are cross-pollinating and some are self-pollinating. In a survey of Swedish blackberries, Nybom (1985) found that the anthers of self-pollinating species burst the day after the petals unfolded and remained spread out for two days to allow cross-pollination. They then arched over the stigmas to facilitate self-pollination. In cross-pollinating species the stamens never touched the stigmas and shrivelled up beneath them. In nature the source of pollen is important when pseudogamy is facultative, but for domesticated types, for example, *R. laciniatus*, the self-pollinating mechanism ensures a good set of drupelets.

Most polyploid blackberries of the Moriferi are either facultative or obligate apomicts. As well as normal sexual embryo sacs, unreduced egg cells arise either from somatic cells by mitotic division (apospory), or from megaspore mother cells following restoration of the unreduced chromosome number. This may occur by the formation of restitution nuclei at the first division of meiosis during which there is a limited amount of crossing over, or by some other process (diplospory). The egg cells' further development depends upon pollination, which apparently stimulates growth through its effects on the endosperm. Clearly, the genetic consequences of apospory and diplospory are very different.

Christen (1950) described how diplosporous embryo sacs arise from generative archesporial cells and aposporous embryo sacs from somatic chalazal cells, often in the same ovule. He considered that meiosis was not involved in the development from archesporal cells, though these cells were clearly capable of meiotic development. He found that *R. caesius* was completely diplosporous, and that *R. vestisus* and *R. thyrsoides* of the *Moriferi* had 62 and 54% of aposporous embryo sacs respectively, together with some normal sexual embryo sacs. Czapik (1981) also found that *R. caesius* had both diplosporous and sexual embryo sacs. Sexual and aposporic embryo sacs were also described for some polyploid North American blackberries by Pratt and Einsett (1955).

Christen's interpretation was challenged by Dowrick (1961, 1966), who considered that both meiosis and embryo sac formation were completely normal in the European species *R. caesius* and *R. calvatus*, which are considered to be obligate apomicts, and *R. laciniatus*, which is considered to be a facultative apomict. He reported that embryo sac formation followed the Polygonum type of development, that there was no evidence of restitution during meiosis and that the embryo sac was haploid. Dowrick claimed support from the observations of Markarian and Olmo (1959) on *R. procerus*, and from those of Christen (1950) on *R. caesius*, but it should be noted that Christen did not regard the diplosporus development that he reported as sexual.

Although Dowrick found that the egg nuclei of these species were haploid, the mechanism for restoring the diploid chromosome number in the absence of fertilization was not discovered until Gerlach (1965) made a histological study of development after pollination in *R. caesius* subspecies *aquaticus*. He found that the reproductive behaviour of this subspecies was plastic, with variants of both sexual and non-sexual reproduction occurring together. Alongside the normal sequence of sexual reproduction, there were instances where two sperm nuclei fertilized the egg nucleus; in some of these the two nuclei originated from the same pollen tube and in some they were from different ones. There was also a non-sexual development which occurred when a pollen tube penetrated a synergid without its contents being released into the embryo sac. This stimulated the egg nucleus to develop abnormally as follows: at the first mitosis the spindle formed at right angles to the long axis of the egg cell instead of parallel with it and along the polarity axis, and it was then followed by fusion of the two daughter nuclei to give a diploid nucleus, instead of by cell wall formation to give cell division. Diploidization without fertilization was thus achieved, and subsequent development was normal.

There were several variations on this process. Sometimes the diploidization process was repeated to give further doubling of the chromosome number. Occasionally, when the synergid burst after diploidization had been stimulated, the diploidized egg cell was fertilized to give an embryo with a preponderance of maternal germplasm. Such a process in different material may have given the Boysenberry and the Youngberry (see p. 63). In yet another variation, the egg cell divided once to give a two-celled haploid proembryo, after which diploidization proceeded either simultaneously to give a diploid embryo or irregularly to give a haploid–diploid chimera.

Endosperm development was normal in these variants, and usually achieved a four- to eight-nucleate stage by the time that the egg began to divide.

Gerlach concluded that the behaviour of the synergid nuclei was crucial. Development without fertilization always occurred when the synergid penetrated by the pollen tube failed to burst. The other crucial factor was the failure to establish a lengthwise polarity in the egg cells. Lengthwise polarity is an essential precondition for normal growth and differentiation in plant tissues, and a precondition for abnormal nuclear fusion leading to diploidization was the positioning of the two haploid daughter nuclei at right angles to the polarity gradient. Thus for *R. caesius* there is a connection, not understood, between failure of the synergids to burst after pollen-tube penetration, initiation of development without fertilization, non-polar orientation at the first division of the egg and failure of cell wall formation.

In some species the genotype of the pollen has considerable influence on the proportion of hybrid to non-hybrid progeny obtained, though it seems certain that the differences observed in segregation reflect differences in survival ability and not in frequency of formation. Higher proportions of hybrid progeny are obtained when *R. laciniatus* is pollinated by diploid blackberry species than when it is pollinated by species of higher ploidy (Dowrick, 1961, 1966). Indeed there is a significant correlation ($r = 0.596$) between the proportion of non-hybrid progeny obtained and the ploidy of the pollen parent (Jennings, *et al.*, 1967), though differences in pollen ploidy do not account for all the differences: for example, the effect of haploid raspberry pollen on segregation is closer to that of diploid blackberry pollen than to that of haploid blackberry pollen. This work with *R. laciniatus* also emphasized the species' low fertility and the low germinability of its seed: the number of seedlings obtained was a fraction of the number of embryos which could have developed. Hybrid embryos were subject to less selective pressure and survived more often when haploid pollen was used.

Another interesting result from this work with *R. laciniatus* was the finding that the progenies contained individuals with chromosome numbers of 14, 21, 28, 38 or 42. The cut-leaf character of *R. laciniatus* shows variable recessiveness depending on the number of recessive alleles present, and so the genetic make-up of these plants could be guessed. This suggested that some of the variations could have been predicted from the diverse reproductive behaviour described by Gerlach. Thus from the cross of tetraploid *R. laciniatus* with diploid *R. idaeus* there were some plants which almost certainly arose from fusion between a reduced egg and haploid pollen, some from fusion of a diploidized egg and haploid pollen and some from fusion of a diploidized egg and either two haploid pollen nuclei or an unreduced one. There were also maternal haploids (Fig. 6.2).

A genetical implication of the diploidizing procedure of blackberries is

Fig. 6.2. Variation in the expression of the recessive cut-leaf characteristic of *Rubus laciniatus*. Leaf (a) is *R. laciniatus* itself ($2n = 28$); leaf (b) is from a diploid segregate of this species ($2n = 14$) and leaves (c) to (f) are from the hybrid progeny of a cross between *R. laciniatus* and "Malling Jewel" raspberry ($2n = 14$), where the chromosome numbers were $2n = 21$, 28, 35 and 42, respectively. The last four probably arose respectively from the fusion of nuclei of a diploid egg and a haploid pollen grain, a diploid egg and a diploid pollen grain, a tetraploid egg and a haploid pollen grain and a tetraploid egg and a diploid pollen grain. (From Jennings *et al.*, 1967.)

that the progeny should be homozygous. However, the species are probably amphidiploids, and the procedure gives homozygosity only between the chromosomes of each of the two complexes which make up the tetraploid. Heterozygosity will be retained as a result of differences between the two complexes, and occasional allosyndetic pairing between chromosomes of the two complexes will lead to segregation. This explains the frequent observation that pseudogamous blackberry progenies are not identical but very similar to the maternal parent. Such progenies are termed "subsexual".

6.7 POLYPLOIDY, PSEUDOGAMY AND DIOECY

In general, all diploid species of *Rubus* are sexual, and polyploid species frequently show either facultative or obligate pseudogamy. Facultative pseudogamy occurs in all the *Rubus* subgenera that have been studied, but it is most common in the blackberries. However, as Gustafsson (1943) points out, species hybrids of blackberries are sexual, even if their parents are facultatively pseudogamous: sexuality is a kind of heterosis phenomenon and pseudogamy is a recessive segregation phenomenon. This gives great evolutionary potential, because recombination in the F1s allows wide segregation in the progenies, from which the more successful recombinants are stabilized within narrow limits following segregation of the subsexual pseudogamous system.

Pseudogamy does not occur in the polyploid Western American blackberries of the *Ursinii*, though *R. vitifolius* of this group can be induced to give non-hybrid progeny (Thomas, 1940*a*). Dioecy is characteristic of this group, and also of the octoploid subgenus *Chamaemorus*. In the *Ursinii*, dioecy is a recessive characteristic and the female is the heterozygous sex, segregating equal numbers of male and female plants (Crane, 1940; Haskell, 1962). There is also evidence that *R. chamaemorus* has female heterogametry. Genetic differentiation of sex occurs in some forms of *R. idaeus* and is associated with two major genes (see p. 168).

7 Diseases Caused by Fungi and Bacteria

Many raspberry and blackberry diseases are caused by fungi and bacteria. It is convenient to group them into diseases of roots, diseases predominantly of canes, diseases predominantly of leaves and canes and diseases of flowers and fruit. Additional references are given by Converse (1966) and Butler and Jones (1949).

7.1 DISEASES OF ROOTS

7.1.1 Verticillium Wilt (*Verticillium albo-atrum* Reinke and Berth. and *Verticillium dahliae* Kleb.)

Verticillium wilt is a serious disease of black raspberries in parts of Oregon, U.S.A. It also occurs in red raspberries and blackberries, particularly in California, but it is rarely serious in Europe, though it has been reported for blackberries in the south of England.

Both *V. albo-atrum* and *V. dahliae* have been reported as the causal agents, though opinions differ as to whether these two forms are distinct species or biotypes of a single species. Strain variation is considerable for both, and not all strains are capable of causing a wilt disease of raspberries. *V. albo-atrum* occurs primarily in the top 30 cm of soil, but it has been recovered down to 90 cm; and in California it has been known to survive for 14 years in the soil in the absence of a known host. The fungal hyphae enter root hairs or penetrate the root cortex directly and make their way to the xylem vessels. They then move throughout the plant and additional spread occurs through conidia in the transpiration stream. The severity of the disease largely depends on the inoculum potential of the soil and this depends on crop history. The strain of the pathogen present is also important, and there are indications that cold temperatures in winter and spring are conducive to severe symptom expression in the following summer. The plant wilts when its transpiration system is too choked to keep pace with water losses, and this problem becomes acute under the stress of hot dry summer weather. It is likely that symptom expression is aggravated by toxins produced by the fungus, because pure-culture filtrates of the fungus also cause wilting and xylem discoloration in raspberries.

In black and purple raspberries the early symptom is a pale leaf colour in mid-summer, followed by a recovery as temperatures drop in autumn. Plants usually die in the first or second season after infection, when first the basal and then the upper leaves turn yellow, wilt and die. The canes become stunted and may take on a blue colour, sometimes in the form of blue stripes, giving the disease its name of blue-stem or blue-stripe wilt. Infected xylem vessels usually become red.

Red raspberries have similar leaf symptoms but they are usually less severe; frequently the leaflets fall before the petioles, and cane discoloration is not as conspicuous as in black raspberries. Differences in susceptibility occur among cultivars but the resistance found is not adequate to provide a control. Indeed, it is not certain whether the resistance shown by a few cultivars is adequate for all strains of the fungus; "Cuthbert", for example, has shown resistance in Oregon and "Latham" has shown resistance in Michigan. Susceptibility is particularly variable among the blackberries and hybrid berries grown in California. Symptoms in these fruits are generally similar to those for raspberry, though the intensity is more varied. Boysenberry and Youngberry, for example, are very susceptible, while Loganberry, "Himalaya Giant", "Evergreen" and the indigenous Pacific coast blackberries are resistant.

In California, soil fumigation prior to planting is an effective control, but it is expensive and is not used widely for raspberries. It is better to avoid infected land, and to avoid spreading the disease by using only healthy certified nursery stock. This is important because the pathogen can survive in nursery stock without symptoms and diseased stock can be planted unwittingly.

7.1.2 Root rots associated with infection with *Phytophthora* species

Several root rot diseases have been associated with infection with *Phytophthora* species. Converse and Schwartze (1968) isolated a species from affected plants in North America and identified it as *P. erythroseptica* var *erythroseptica* Pethyb. This isolate killed plants of "Canby" but not "Newburgh". Seemüller *et al.* (1986) report that severe outbreaks of root rot and raspberry die-back in Germany were associated with isolates with similar characteristics to this species. Duncan *et al.* (1987) described a survey of root rot outbreaks in Britain in which a group of isolates of *P. megasperma* Drechsler pathogenic to raspberries were associated with the most severe instances. In these outbreaks; areas of dead or dying plants increased in size as the disease spread, while the surviving plants had only a few new canes which showed brown-black lesions at their bases. Removal of their bark showed that the woody tissue was discoloured with distinct margins to the affected areas, and any laterals that emerged usually wilted before their leaves expanded fully. The older roots also showed necrosis and there were very few feeder roots. Duncan also isolated *P. syringae* (Kleb.) Kleb., *P. drechsleri* Tucker, *P. cactorum* (Lebert & Cohn) Schröter and *P. cambivora* (Petri) Buis. from diseased plants, but in pathogenicity tests they found that these species caused only moderate amounts of root rot and no aerial symptoms, except for *P. cambivora* which caused a mild yellowing and wilting of the lower leaves. These symptoms were more severe if water-logging occurred near the time of inoculation, and for all isolates, wet, poorly drained conditions exacerbated the root rot symptoms. Duncan *et al.* (1987) also concluded that the North American and German isolates previously described as *P. erythroseptica* should be assigned to this group of *P. megasperma* isolates.

Barritt *et al.* (1979) tested many clones in a field in Washington, U.S.A., where *Phytophthora* species were thought to be active, and found that "Latham", "Newburgh", "Cherokee", "Pathfinder" and "Sunrise" and some of their derivatives were resistant, while "Lloyd George" and many of its derivatives were susceptible. Progeny tests showed that the resistance was highly heritable and both "Latham" and "Newburgh" gave high percentages of resistant progeny. "Latham" and derivatives of *R. spectabilis* were the most resistant in pot tests in Scotland.

Root rot diseases of raspberry are frequently associated with heavy soils in western North America. They have similar symptoms to those found in Europe, but it is probable that the primary pathogens involved have not all been identified. Apart from *Phytophthora* species, the fungi most commonly isolated from affected roots include *Fusarium* sp., *Cylindrocarpon radicicola* (McAlp), *Pythium* sp., *Leptosphaeria coniothyrium* (Fuckel) Sacc., *Phoma* sp. and *Rhizoctonia* sp. Some of these may interact with the nematode *Pratylenchus penetrans*. Both in North America and Europe there is evidence that root rot diseases are spread in infected planting material.

7.1.3 Root rots associated with land previously occupied by trees

Root rots of *Rubus* species caused by *Armillaria mellea* sometimes occur on land previously occupied with trees, and affected plants invariably die after a period of wilting. Rhizomorphs of the fungus spread in widening circles from the point of infection and it is necessary to dig all the roots from a ring around the plants to achieve control. Silver leaf disease (*Chondrostereum pur-*

pureum (Pers: Fr) Pouzar), which infects canes, invades the roots and kills whole plants, and a white root rot disease associated with a species of *Vararia* (Pascoe *et al.*, 1984) are locally important on recently cleared land in Australasia.

7.2 DISEASES PREDOMINANTLY OF CANES

7.2.1 Cane blight (*Leptosphaeria coniothyrium* (Fuckel) Sacc.)

Cane blight is caused by infection of raspberry cane wounds by *L. coniothyrium* and is a serious disease in Europe and New Zealand, but, although it occurs in North America, it is rarely of economic importance there unless irrigation is used, because spread of the fungus is usually limited by low rainfall in July (Williamson *et al.*, 1986). Cane blight is essentially a disease of wounds, and may occur for example in wounds inflicted by machines, hoes, pruning, training wires or by primocanes rubbing against the sharp bases of old fruiting canes. Wounds are caused by the catching plates and vibrating fingers of some machine harvesters and an increased incidence of cane blight has consequently been a serious drawback of machine harvesting in Europe (Williamson and Hargreaves, 1978).

Inoculum is provided by fruiting bodies on old cane stubs and on wounds of fruiting canes. Within two weeks of infection, wounds on green primocanes show dark-brown lesions which become white during autumn and contrast with the natural autumn colours of the cane. The lesions may show black subepidermal pycnidia covered with sooty spores in March. Scraping a cane's surface to remove the epidermis, cortex and periderm reveals brown vascular "stripe" lesions which girdle and may totally occlude the vascular cylinder, thereby blocking translocation to the sector supplying affected buds (Fig. 7.2). Buds immediately above the lesions usually fail to produce fruiting laterals, or the laterals which emerge from them may be retarded and wilt at any time from leafing-out to harvest; death of the cane above the point of infection is also common, especially in infections which occur in early summer. In a bad year in Scotland, the loss in yield in the year after machine harvesting was estimated to be about 30% and attributed to infection of wounds inflicted by the harvesters (Hargreaves and Williamson, 1978).

Since the pathogen can only infect wounds, control of the disease is achieved by reducing all causes of cane damage. For example, harvesters can be designed so that the catching plates of the machines rub the canes so lightly that no more than the surface bloom is removed (Williamson and Ramsay, 1984); similarly, in "Glen Clova", cane vigour control through chemical burning can reduce the incidence of vascular lesions from 28% in untreated canes to 2% in the replacement canes, simply because the later are shorter and less prone to picker and wind damage (Williamson *et al.*, 1979). Pre- and post-harvest sprays with a systemic fungicide are also effective.

Breeding for resistance is a long-term possibility. For any cultivar, resistance builds up during the growing season: it is low in July, but is sufficiently augmented by August that new infections cause only small lesions which do not girdle the cane and rarely cause bud failure. Resistance builds up more quickly in "Latham" and in some asiatic species of the *Idaeobatae* (Jennings, 1979a).

7.2.2 Midge blight

Midge blight occurs only in Europe. It is a disease complex resulting from invasion of the feeding wounds of the cane midge (*Resseliella theobaldi* (Barnes)) by several pathogens. Grünwald and Seemüller (19779) detected cellulase and esterase activity in the salivary glands of the midge larvae, and showed that this enabled the larvae to degrade the cell wall constituents,

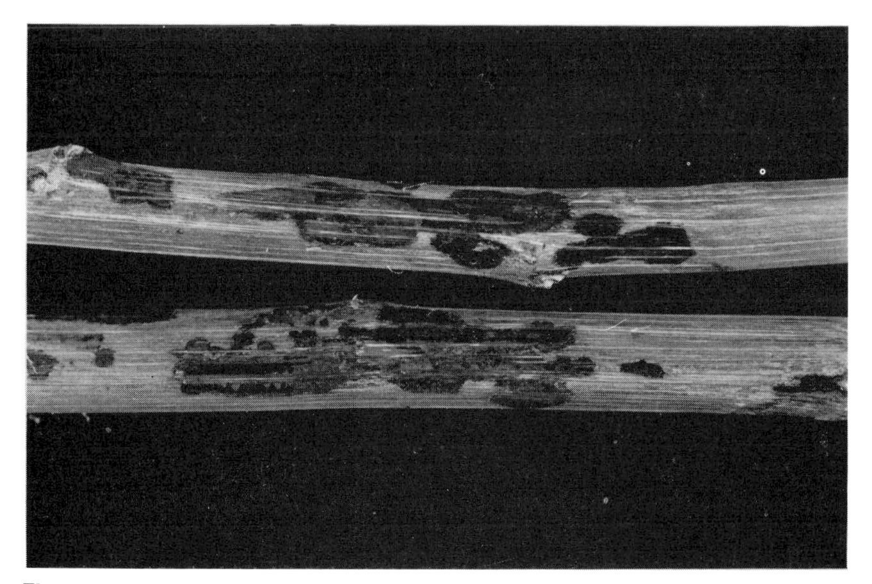

Fig. 7.1. Brown, irregular "patch" lesions revealed by scraping canes affected by midge blight. The lesions are associated with the boundaries of feeding areas of the raspberry cane midge (*Resseliella theobaldi*) and are usually infected with several fungal pathogens.

especially suberin, and to deprive the periderm of its protective properties, thereby allowing wound pathogens to penetrate to the internal cane tissues and cause serious damage. Pitcher and Webb (1952) consistently found *Didymella applanata* and *Fusarium culmorum* among the pathogens isolateed from the midge blight complex and *Leptosphaeria coniothyrium* more rarely, though they commented that the latter caused rather more damage than the other pathogens. They found that the relative frequency of pathogens varied between years and localities. Other workers have isolated other pathogens, but the situation was clarified by Williamson and Hargreaves (1979), who recognized two types of vascular lesion when canes were scraped to expose the vascular cylinder (Figs. 7.1 and 7.2). One type, termed a "patch" lesion, was brown and irregular in outline and occurred within the precise boundaries of the midge feeding areas; such lesions could cover up to 20% of the stele surface at the bases of the canes before yield was reduced. The other type, also brown, was termed a "stripe" lesion. This type spread proximally and distally from the infection points during winter. Isolations from within the vascular tissues of patch lesions produced principally *Fusarium avenaceum* (Fr.) Sacc., various species of *Phoma* and the Phoma stages of *Didymella applanata*, and isolations from "stripe" lesions adjacent to the patches yielded *L. coniothyrium* and *F. avenaceum*. Clear, though indirect evidence for the involvement of *L. coniothyrium* in the

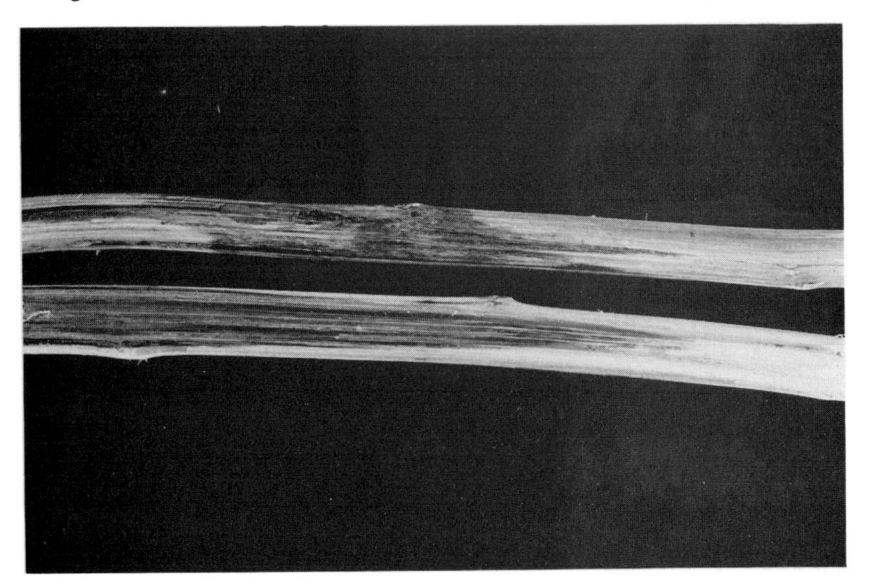

Fig. 7.2. Elongated, vascular "stripe" lesions revealed by scraping canes affected by cane blight. The lesions spread proximally and distally from points of infection with *Leptosphaeria coniothyrium*.

midge blight complex was later provided by the finding that insecticide sprays considerably reduced the incidence of both patch and stripe lesions (Williamson, 1987).

Midge blight becomes a serious disease when raspberry canes develop extensive natural splits (see p. 145) during the egg laying periods of first- and second-generation midges. "Glen Clova" is particularly prone to do this. Cane losses of from 50–90% have been recorded in Scotland, but they are greatly reduced by cane vigour control treatments which chemically burn the first flush of young canes when it is 10–20 cm high. This is because the replacement canes which grow after the treatment are later and have developed few splits by the time that the first-generation midges are emerging. The midges are consequently unable to find sites for egg laying; cane splitting is in fact both delayed and much reduced in extent by this treatment (Williamson *et al.*, 1979). Similarly, resistance to midge can be achieved through selection of types whose canes either tend to split late in the season, or have a capacity to seal their splits (see Fig. 11.2).

7.2.3 Cane spot (*Elsinoë veneta* (Burkh.) Jenkins)

Raspberry cane spot, or anthracnose as it is more commonly known in North America, is an extremely serious disease of black and purple raspberries and susceptible cultivars of red raspberries (Fig. 7.3). It occurs on some blackberries and can be severe on Loganberries and Tayberries. The symptoms can occur on canes, leaves, flower buds or fruits, but the most serious are those on the canes. On black or purple raspberries the first symptoms are small, slightly sunken purple spots on young canes; these enlarge and their centres change to a grey colour, leaving the margins purple and sometimes slightly raised. The older lesions are deep and extend to the centre of the canes, causing them to dry out and develop cracks several centimetres long. In severe cases the canes assume a warted or knotted appearance because of tissue swelling beneath affected areas. Canes that have been girdled die during the winter, while buds near the cankers either die or give irregular development of fruiting laterals. A die-back from the tip is also common.

On canes of most red raspberry cultivars the lesions are similar but smaller, less frequent and more greyish than those on black or purple raspberries. The greying of the bark is so conspicuous that the disease is sometimes referred to as "grey bark". The bark also has reddish pimple-like acervuli arranged in concentric circles.

Leaf-spot symptoms of this disease occur on all kinds of raspberry. They are usually small but may reach 1.5 mm in diameter and are recognized by their conspicuous purple margin and light grey centre; this may drop out and

Fig. 7.3. Raspberry fruiting canes infected with (left to right) *Didymella applanata* (spur blight), showing small black pycnidia and perithecia on a "silvered" cane; *Botrytis cinerea* (cane botrytis), showing blister-like sclerotia on a "silvered" cane; and *Elsinoë veneta* (cane spot), showing sunken greyish spots.

leave a shot-hole effect, and in severe cases the leaves may curl and drop prematurely. Petioles, peduncles and fruit pedicels and even the fruit are also subject to attack; the girdling of the pedicels can seriously reduce yield and rusty-brown scabby lesions on the fruit result in loss of quality. Alternatively, affected drupelets may remain green so that the fruit is deformed or the entire fruit may fail to mature.

In contrast to most other cane pathogens, *E. veneta* attacks succulent young growth rather than mature tissues. The canes are most susceptible when they are 10–30 cm high, and hence infections after July are not so serious as earlier ones, and lesions formed from August infections remain small and shallow. In Scotland, the critical months for infection are May, June and July, when acervuli at the centre of the deep, ash-grey lesions produce masses of conidia for dispersal by rain splash on to the young canes. Both conidia and ascospores are produced on over-wintered canes to coincide with their leafing out, though ascospores are relatively rare. The invading fungus kills the tissues of the cambium and phloem and is not impeded by the periderm like other cane pathogens.

The fungicide sprays used to control botrytis fruit rot give some control of cane spot, but the sprays should be applied earlier to coincide with spore dispersal in severely affected plantations.

Large differences in resistance occur among red raspberry cultivars and some important cultivars, notably "Glen Clova", "Skeena", "Glen Moy" and "Leo" are highly susceptible. There is evidence that a major dominant gene is involved in resistance and that gene *H* reduces resistance (Jennings and McGregor, 1987). Related Asiatic species such as *R. coreanus* of subgenus *Idaeobatus* are also good sources of resistance (Williams, 1950; Keep *et al.*, 1977b).

7.2.4 Purple blotch of blackberries (*Septocyta ruborum* (Lib.) Petrak)

Purple blotch disease of blackberries (Fig. 7.4) produces symptoms with some similarity to those of cane spot, but it is limited to the stems of blackberries and does not attack leaves. It occurs throughout Europe but not in North America. The fungus invades the stomates of the primocanes during the summer and penetrates the parenchyma and collenchyma, but exposure to cold is needed for symptom expression and so symptoms do not appear until the following spring. They start as red pinhead spots which increase in size, fuse into large purple blotches and may cover entire sections of the stem. Rows of raised black pycnidia may be present. If the disease is severe, the leaves and flower buds wilt and shrivel, and the stem itself may shrivel up as if it had suffered frost damage (Koellreuter, 1950; Oort, 1952).

7.2.5 Spur blight (*Didymella applanata* (Niessl) Sacc.)

Spur blight (Fig. 7.3) is a disease of individual red raspberry nodes caused by *Didymella applanata.* It affects red raspberries throughout Europe and North America, but is rarely found on black raspberries or blackberries, though it can occur on Loganberries. The disease occurs first as small brown

Fig. 7.4. Lesion of purple blotch on a blackberry fruiting cane caused by infection with *Septocyta ruborum*.

lesions at leaf margins and near the large leaf veins, often separated from green tissue by a yellow band. The leaves may eventually develop a brown shrivelled appearance. The leaves are the main infection sites, and although the fungus has been reported to pass from the petiole to bud scales in Canada, this has not been found in Scotland. Histological study of Scottish material showed that the infection route to the axillary buds is blocked by an outer suberized phellem layer of suberized and lignified cells across the adaxial cortex of the petiole, but that the fungus can freely invade the outer tissues of the cane beneath the node (Williamson, 1984). Hence the fungus grows basipetally within the petioles and colonizes the cortex of the cane around the buds and the undersides of the swollen leaf bases, where it kills the cells and induces a chestnut-brown lesion. Symptom development has been attributed to the production of an exocellular phytotoxic glycopeptide (van Broekhoven *et al.*, 1975).

Although ascospores are released in early summer, they probably play a minor role in infection because the young canes are relatively resistant at this time. Pycnospores become abundant from mid-June to mid-August and probably have the major role in infecting the leaves, which become highly susceptible as they begin to senesce when light is excluded from the cane bases. Since canes become too mature to be infected after August the period for infection to occur is limited. Indeed, although moist conditions are required for infection, it is likely that disease spread in Britain is more often limited by periods of plant susceptibility than by the occurrence of rain or dew, or by the availability of inoculum (Burchill and Beever, 1975). And it is probably because the replacement canes produced in response to cane vigour control are physiologically juvenile that they tend to show few spur blight infections (Williamson *et al.*, 1979). However, the susceptibility of canes is considerably increased by a period of relatively high temperature before attack and this factor contributes to the occurrence of symptoms at nodes over a high proportion of the canes' length in western Canada (Pepin and Williamson, 1986).

By July or August the lower half of primocanes may be covered with chestnut-brown discoloured spur blight lesions at the nodes. These become silvery grey in winter and may extend to internodal regions; they can be recognized by the presence of masses of pycnidia and perithecia. The buds are rarely killed because they are not invaded by the fungus, but they are often so weakened and reduced in size that they fail to develop when in competition with uninfected buds. They are more prone to be killed by frost, however (Rebandel *et al.*, 1985). The effect of the disease on bud competition seems to vary and is relatively less in a vigorous though susceptible cultivar like "Glen Clova" (Williamson and Dale, 1983; Pepin *et al.*, 1985). Moreover, the effect on yield is not proportionally high, because raspberry canes have a strong capacity to compensate for failure at some fruiting nodes by increased production from others (Braun and Garth, 1984).

Red raspberry cultivars differ widely in resistance to the pathogen (Daubeny and Pepin, 1975; Jennings, 1982a), and cane morphology has a considerable effect on resistance; hairy canes (gene *H*) tend to be less affected, especially if they are also spine-free, and so do canes with a dense waxy bloom. The reasons for this are not known: it was first thought that the presence of hairs aided escape because it promoted water run-off (Jennings, 1962), but this is not the main reason because hairy canes are also more resistant to mycelial inoculation of wounds (Jennings, 1982b). Several

Asiatic species (see Fig. 7.3) show resistance and can be used in breeding (Keep *et al.*, 1977b; Jennings, 1982a).

7.2.6 Cane Botrytis (*Botrytis cinerea* Pers: Fr.)

This pathogen is best known as the causal organism of fruit grey mould. It also infects the leaves and petioles in late summer in much the same way as *D. applanata* does, though its subsequent spread along the cane can be much more extensive. The invading fungus reaches the canes and forms tan coloured lesions around the buds, often surrounded by a characteristic "watermark" symptom caused by changing growth rates of the fungus in fluctuating environmental conditions. The lesions become white during winter and develop black blister-like sclerotia which release spores during the spring (Fig. 7.5). The cane lesions are the principal initial sources of inoculum for infection of flowers and fruit.

The effect of the disease on bud development is similar to that of spur blight, and "Glen Clova" is similarly tolerant. Indeed, similarities with spur blight are notable for most aspects of the disease: other examples are the relative resistance of juvenile canes and of the replacement canes produced in response to treatments applied for cane vigour control (Williamson *et al.*, 1979); there are also similar effects on the disease of cane hairiness and waxy

Fig. 7.5. Fruits of *Rubus pileatus*, an asiatic species used as a source of genes for resistances to several fungal pathogens.

bloom and in Scottish studies it was observed that resistance to cane botrytis and spur blight behaved as a single character in progenies derived from the red raspberry "Chief", *R. occidentalis*, *R. pileatus* or *R. coreanus* (Jennings, 1983). It also behaved in this way in studies of resistance in seven raspberry cultivars (Williamson & Jennings, 1986).

7.2.7 Blackberry cane canker (*Botryosphaeria dothidea* (Moug.: Fr.) Ces. & de Not.

Cane canker has become a serious disease of the new spine-free blackberry cultivars grown in eastern United States. *B. dothidea* establishes itself saprophytically in senescing petioles and then invades healthy cane tissue near the buds, where it stimulates reddish-brown cankers to develop and girdle the stem. This causes the cane above the canker to wilt and die at about the time that the fruit is ripening (Maas, 1986).

7.2.8 Crown gall and stem gall—A bacterial disease (*Agrobacterium radiobacter* var. *tumefaciens* (Smith & Townsend) Com.

The taxonomic status of the bacteria isolated from crown and stem galls of raspberries and blackberries have been controversial since the first isolate was made in 1907. Many authors consider that distinct species of bacteria, *Agrobacterium tumefaciens* and *A. rubi*, cause crown gall and stem gall respectively. But the range of variation in these bacterial isolates is very great, and as more of them have been studied the differential physiological and pathological characteristics proposed by Hildebrand (1940) for separating the two species have proved more and more unsatisfactory (McKeen, 1954; Keane *et al.*, 1970). Keane *et al.* (1970) therefore recognized only one causal agent, *A. radiobacter*, which they subdivided into the biotypes *tumefaciens* and *rhizogenes*. However, forms intermediate between these two biotypes have since been found, especially in *Rubus* and *Vitis*, and it appears that some strains produce galls on fruiting canes, crowns and roots, while others produce them only on the crowns and roots. The crown-gall form produces galls on inoculated canes but seems unable to move up the canes from the root zone (McKeen, 1954). Until the taxonomic relationships are clarified the name *Agrobacterium radiobacter* var. *tumefaciens* seems appropriate for all forms of root, crown and cane gall. It can be regarded as synonymous with *A. tumefaciens*, *A. rubi*, *Bacterium tumefaciens*, *Pseudomonas tumefaciens*, *Bacillus tumefaciens* and *Phytomonas tumefaciens*. The organism has a wide host range besides raspberries and blackberries.

Fig. 7.6. Crown galls at the base of a raspberry cane caused by infection with *Agrobacterium radiobacter* var. *tumefaciens*.

Gall diseases (Fig. 7.6) occur on raspberries and blackberries wherever they are grown. The causal bacteria are essentially pathogens of wounds or cuts, and hence they are most likely to infect plants during their propagation. A common symptom is a spongy or hard gall outgrowth at the site of a cut made during propagation, but cracks in the stems induced by winter frosts, root-attacking insects or implements are also potential entry sites. The size of the galls varies considerably; they may be several inches in diameter or the size of a small pea. In raspberries and blackberries galls occur most commonly on roots and occasionally on stems and crowns; but in black raspberries they occur mostly on the stem region at or just below soil level; hence the name "crown gall".

On blackberries and hybrid berries the stem lesions are elongate and may extend along the whole length of a cane. In a typical outbreak irregular masses of tissues balloon out at points along the canes to form galls; they appear first in May as small protuberances and become progressively larger and eventually turn brown and die in the autumn. Continuous internal pressures within the canes cause them to rupture and split. If the symptoms are severe the canes may show a leaf wilt while the fruit is developing, and the berries may consequently fail to ripen and tend to dry up.

The disease is most severe in cool wet soils of high pH. After infection, the host tissues are stimulated to make localized growth and produce galls around the wounds. The bacteria are apparently confined to outer parts of the galls, and these are constantly being sloughed off into the soil where they are reputed to survive for many years. In spite of the alarming symptoms, vigorous infected raspberry plants have a remarkable capacity to recover, and plants showing severe infection in one year may be completely free of symptoms in subsequent years.

The disease is controlled by planting healthy stock, and by avoiding infested fields and close cultivation likely to cause plant wounds. Some avirulent strains of *A. radiobacter* produce a bacteriocin which acts as an effective antibiotic against many virulent *Agrobacteria* strains and can be used to protect plants against them. However, no effective bacteriocin-producing strain has been found to protect against all the strains which attack raspberries and blackberries, and the protection afforded depends on the strains present.

Cultivars vary in susceptibility: "Glen Clova" and "Malling Delight" are particularly susceptible (Swait, 1980) while "Willamette" has a useful degree of resistance (Zurowski *et al.*, 1985).

7.2.9 Leafy gall (*Corynebacterium fascians* (Tilford))

A disease known as leafy gall is recognized by the mass of short shoots produced from the roots at or near ground level. The shoots are frequently distorted and thickened, and are formed from root buds which are normally dormant but are stimulated into growth by the causal bacterium, which may also stimulate the initiation of new buds. The bacterium is *Corynebacterium fascians*, which persists in the soil and has a wide host range.

7.2.10 Fire blight (*Erwinia amylovora* (Burr.))

Fire blight is a rare disease of raspberries and blackberries which causes death of the stem tips, leaves and flowers and produces mummified fruit.

The strains of *E. amylovora* involved are distinct from those that infect the apple and pear.

7.3 DISEASES PREDOMINANTLY OF LEAVES AND CANES

Orange rust (*Gymnoconia peckiana* (Howe) Trott.)

Orange rust is important only in North America, where it is the most important of the rusts; it attacks blackberries and black raspberries but not red raspberries nor some of the hybrid berries such as Boysenberry. It attacks the leaves as they unfold in spring and continues to develop on new leaves as they are produced.

The plants usually become systemically infected through local infections in the young shoots in autumn. Small brown telial spots develop on infected leaflets, and teliospores from them produce sporidia that infect the buds at the cane tips; in trailing blackberries mycelium from the infections follows these tips as they root at the end of the season and enters the new root system. In black raspberries, where the cane tips are usually weakened too much to root, the fungus grows into the main stem at the base of the affected shoot and enters newly formed roots from there. For both kinds of fruit, the young shoots produced in the years after infection are spindly, stunted and bushy and produce a completely non-fruitful witch's broom type of growth. The new shoots of black raspberries are particularly weak, being free of spines and having small yellow leaves which are susceptible to mildew. The young leaves on these shoots are first covered with minute reddish pycnia and later with blister-like orange aecidial postules on their undersides. Recovery may occur later with no symptoms visible after late July, though the plants remain stunted. In blackberries there is a microcyclic form of *G. peckiana* known as *Kunkelia nitens* which lacks teliospores, its aeciospores taking their function.

Infected plants produce orange rust postules each spring for as long as they live, and the main control measure is to make sure that a rust-free nursery stock is used. Plants should be inspected a month after planting and whole plants including the roots rogued if they show infection.

7.3.2 Cane and leaf rust (*Kuehneola uredinis* (Lk.) Arth.)

Cane and leaf rust, alternatively known as yellow rust or blackberry rust, is a serious disease of some blackberries in certain parts of North America, notably the south-east states and the Pacific north-west. It is not systematic, but it is easily mistaken for orange rust. It rarely attacks raspberries. The first symptoms are lemon-yellow uredinia which split the bark of the fruiting canes in spring. Urediniospores from these infect the leaves of the fruiting canes and may lead to premature defoliation in serious outbreaks. Telia develop later and infect primocane growth, where the sexual stages develop in October–November on the lower leaves.

7.3.3 Yellow rust (*Phragmidium rubi-idaei* (D.C.) Karst. Syn. *P. imitans* Arth.) and blackberry rust (P. violaceum (Schultz) Winter)

Yellow rust is relatively unimportant because it is only locally and sporadically severe, even though it occurs on red raspberries worldwide. It occurs for example in the Pacific north-west, Australasia and in parts of Europe. Cultivar susceptibility varies widely. A related rust *P. violaceum* (S. F. Schultz) attacks blackberries in Australia, Europe and South America, but not in North America.

Initial infections of raspberry leaves and emerging shoots occur from over-wintered teliospores. These give rise to orange-yellow aecia on the upper surfaces of the leaves in spring, followed by orange to pale yellow uredinia on their under surfaces and on canes in June. Pustules of the latter contain black telia from mid-July to winter. Premature defoliation may result from heavy rust attack early in the season, and over-wintering cane lesions may also become deep and cause the cane to be unproductive in the following summer. In Britain, the disease has become prevalent with the increased growing of "Glen Clova" and "Malling Delight". Two kinds of resistance are known: strong resistance conferred by the major gene *Yr*, which prevents sporulation and occurs in the cultivars "Chief" and "Boyne", and resistance of the "slow rusting" type which is conferred by minor genes and causes a delay in the appearance of pustules and a reduction in their number and size (Anthony *et al.*, 1986). In some germplasm resistance to yellow rust is highly correlated with resistance to cane spot (Jennings and McGregor, 1987).

The effects of *P. violaceum* are so severe in Australia and Chile that the disease has been considered as a possible means of biological control of spreading blackberry hedgerows. This is because the disease is particularly severe on the common hedgerow species, notably *R. procerus* and *R. ulmifolus*, while cultivars of the more important cultivated species, notably *R. ursinus*, are resistant (Bruzzese and Hasan, 1986). Cane lesions near the cane bases can serve as entry points for *Leptosphaeria coniothyrium*.

7.3.4 Late leaf rust (*Pucciniastrum americanum* (Farl.) Arth.)

Late leaf rust, sometimes known as autumn rust or late yellow rust, is another rust of local importance which attacks only red and purple raspberries and occurs in California and the northern parts of central and eastern America. The disease does not normally appear before harvest when mature leaves develop small spots which are first yellow and then brown and have uredinia on their under surfaces. In severe cases the disease causes premature defoliation and renders the canes prone to winter injury. The fruit of late primocane-fruiting cultivars do not escape and some 30% of fruits from "Heritage" were reported unfit for sale in an outbreak from Ohio (Ellis & Ellett, 1981).

7.3.5 Leaf spot (*Sphaerulina rubi* (Demaree & Wilcox))

Raspberry leaf spot, sometimes called Septoria leaf spot, occurs in most parts of North America, but is serious only in south eastern parts, particularly at the southern limits of raspberry growing. The leaf spots occur first on young leaves, where they are small, circular to angular and greenish black in colour; they enlarge and become grey as the leaves mature and may drop out to give a shot-hole effect. The young leaves produce pycnidia whose spore masses are dispersed by rain splash. Perithecia develop in autumn and release their ascospores in spring. Premature defoliation may result, predisposing the canes to winter injury. Small lesions may also occur towards the bases of the canes and may be necrotic.

There is confusion about the identity of the pathogen, but *Rhabdospora rubi* Ell., *Septoria darrowi* Zeller and *Septoria rubi* West can probably all be regarded as synonyms of *Sphaerulina rubi*. Different forms of the pathogen exist and the forms that infect raspberries and blackberries do not cross-infect in the field. Many blackberries are resistant however, as are many Asiatic species which have been used as donors of resistance in breeding. The disease is usually controlled by good plantation hygiene and by the sprays used to control cane spot.

7.3.6 Powdery mildew (*Sphaerotheca macularis* (Wallr: Fr.) Lind)

Powdery mildew of red, black and purple raspberries occurs in North America and Europe. Blackberries and hybrid berries are rarely affected. The disease is most serious in hot dry weather; it appears first in May or June on young leaves and tender stems, producing light green blotches on the upper surfaces, matched on the lower surfaces by white patches covered with powdery masses of conidia. The leaf edges curl; the shoot tips become covered with mealy mycelial growth, produce small leaves, become long and spindly and may rosette prematurely. This causes the plants to become stunted. Flower buds and fruit are also seriously affected: the fungus may prevent late buds from developing into fruit and may render the fruit worthless by completely covering it with mealy mycelial growth.

Cultivars differ widely in resistance (Keep, 1968b; Keep *et al.*, 1977b,c) and resistance in the leaf is not closely correlated with resistance in the fruit. Cultivars like "Puyallup" and "Glen Clova" are highly susceptible and should be avoided in areas where the disease is persistently serious. In general, fruit with tough skin is relatively resistant. Sprays applied to control botrytis fruit rot normally reduce this disease to some extent as well but sprays specific to mildew are better.

7.3.7 Downy mildew (*Peronospora sparsa* Berk. (Syn. *P. rubi* Rabenh.)

Downy mildew can occur on some raspberries, "Marcy", for example, but it is usually associated with blackberries and hybrid berries grown in areas of high rainfall such as New Zealand, where it is severe on Boysenberry and Youngberry. It causes a severe leaf blotch and leads to an early loss of leaves and also becomes systemic. Its most serious symptom is the so-called dry-berry disease, where the receptacles split and the drupelets become dry and hard and render the fruit worthless (Tate, 1981).

7.3.8 Pseudomonas blight (*Pseudomonas syringae* (van Hall))

This bacterial blight has been reported on raspberries in British Columbia. The causal pathogen induces brown water-soaked spots on the leaves, petioles, young primocanes and developing laterals. Young girdled canes usually shrivel, crack and later die, or they may survive and produce weak laterals with small fruit which shrivel and fail to mature (Pepin *et al.*, 1967).

7.4 DISEASES OF FLOWERS AND FRUIT

7.4.1 Stamen blight (*Haplosphaeria deformans* (Syd.))

Stamen blight, known in British Columbia as anther and stigma blight, is a disease of the stamens in which the anthers become filled with a powdery white mass of spores resembling mildew. It attacks raspberries and blackberries in Europe and western North America and has been known to build up into severe epidemics in favourable conditions. The affected stamens produce no pollen, and fruit from affected flowers tend to be malformed, uneven ripening and difficult to pick. The petals have a slightly flattened appearance which makes the flowers look bigger than normal.

Infection occurs in the axillary buds of the primocanes and the mycelium keeps pace with the growth of the fruiting lateral in the following year until it eventually sporulates in the stamens. The disease may therefore be eradicated by removing all the canes and sacrificing a year's crop, but if the outbreak is not too severe the sprays applied against botrytis fruit rot provide an effective control, provided that the spray is set to reach the axillary buds. The planting of certified stocks where prevention of flowering is a requirement of the certification scheme is the best control.

7.4.2 Double blossom of blackberries (*Cercosporella rubi* (Wint.) *Plakidas*)

Double blossom, or rosette, is a disease of major importance on blackberries, but it has only been reported from the more southern states of North America and not from the Pacific Coast. There are very few reports of raspberries being infected. Axillary buds of primocanes are infected in early summer; the fungus occupies the spaces between the bud scales but makes no further growth in the first year unless the buds are forced into growth during a warm autumn. The mycelium increases within the buds during the following winter but does not invade the bud primordia, though it induces bud proliferation. The first symptoms normally occur in the spring after infection occurred. Masses of short leafy shoots then develop and form rosettes or witches brooms at the infected nodes. Infected flower buds are enlarged and malformed and have large leaf-like and often reddened sepals. The styles are elongated and both styles and stamens become covered with white fungal spores. No fruit develops. In some cultivars the fungus invades the stem and may enter the crown of a rooted stem tip.

Cultivars differ in susceptibility: "Himalaya Giant" is immune. Since spores are formed only in the flowers, control of the disease is achieved by similar measures to those used to control stamen blight. Regular sprays during the flowering period are also effective.

7.4.3 Fruit rots

Fruit rot is caused by several fungi, of which grey mould (*Botrytis cinerea* Pers. : Fr.) is easily the most important. A number of other species including species of *Rhizopus*, *Mucor*, *Cladosporium*, *Penicillium*, *Alternaria*, *Fusarium* and *Aureobasidium* play lesser roles. The rapid development of post-harvest fruit rot shortens the shelf-life of fruit and is one of the principal factors restricting fresh-fruit sales in markets distant to centres of production. It is especially common when wet or humid weather occurs during flowering or fruit ripening. In Canada, some 64 to 93% of fruit were found to develop fruit rot between 40 and 60 h after harvest, even though the incidence in the field was less than one per cent (Freeman, 1965).

Fruit affected by grey mould, sometimes known as botrytis fruit rot, show a watery soft rot covered by a grey-brown mass of felt-like hyphae. Spores from over-wintering sclerotia initiate the infection cycle by infecting the flower parts. Many conidia alight on the styles and hyphae from them grow down the stigmas alongside pollen tubes. They reach the ovary and form an endogenous mycelium, while mycelium also persists in the senescing stylar tissues and stamens and provides a further source of inoculum to attack the surface of the ripe fruit (McNicol *et al.*, 1985; Williamson *et al.*, 1987). The importance of the endogenous mycelium for the initiation of grey mould is not yet fully understood, but it is likely that many other infections occur from invasion of the fruit surfaces, especially through small wounds inflicted at picking time. Hence cultivars with a tough skin and firm texture show a degree of resistance (Jennings and Carmichael, 1975a; Pepin and MacPherson, 1980; Knight, 1980a). The cultivars "Nootka" and "Chilcotin" have consistently shown resistance, and good sources of resistance are black raspberry cultivars and *R. crataegifolius* (Daubeny and Pepin, 1981; Barritt and Torre, 1980; Knight, 1980a,b). The discovery that the incidence of fruit rot is considerably increased by inoculating flowers, developing fruit or ripe fruit with dry spores emphasizes the importance of measures to remove rotting material from the plantation, and also the value of cane resistance to *B. cinerea* (Williamson *et al.*, 1987).

A high incidence of grey mould is associated with long intervals between picking dates and a high rainfall during any 10 day period of the picking season. It can be reduced by picking only sound fruit, by picking at an early stage of maturity and by refrigeration as soon as possible after picking under conditions which facilitate air movement around the fruit. Careful handling to avoid bruising the fruit is important.

Infection by *Rhizopus stolonifer* (Ehr ex. Fr.) Lind or by other fungi of the *Mucorales*, such as *Mucor hiemalis* or *M. mucedo*, causes a soft fruit rot which occurs most frequently in post-harvest samples. In some situations these rots may be as important as botrytis rot, and they may be more prevalent when *B. cinerea* is controlled by sprays which do not control fungi of the *Mucorales* (Mason and Dennis, 1978). They produce a dense white hairy mycelium covered with small black pin-head sporangiophores, and are particularly prevalent if storage temperatures are high (e.g. 20°C), and in samples picked late in the season. Resistance to *Rhizopus* fruit rot is also favoured by firm fruit texture (Daubeny *et al.*, 1980).

Moulds caused by species of *Cladosporium*, including *C. herbarum*, *C. cladosporioides* and *C. epiphyllum*, are also common on over-ripe or bruised harvested fruit of red or black raspberries. These fungi produce an olive-green mould, usually on the inside of the fruit and usually several days after harvest. Other post-harvest fruit rots include blue mould, caused by species of *Penicillium*, whose fungal growth is first white and then blue-green; and *Alternaria* rot, caused by *Alternaria* species including *A. humicola*, which has a dark grey mycelium and is common on black raspberries. *Aureobasidium pullulans* and *Fusarium* species are less common fruit rot pathogens. There is a tendency for resistance to botrytis mould to be positively correlated with resistance to *Cladosporium* and *Alternaria* moulds (Knight, 1980a).

8 Diseases Caused by Viruses, Mycoplasma-like Organisms and Genetic Disorders

Altogether, some 26 distinct viruses and virus-like diseases have been reported for *Rubus* crops throughout the world (Jones, 1986a), and they are conveniently classified into four groups according to their mode of transmission. These are the viruses transmitted by the large aphids (*Amphorophora* sp.), the small aphids (*Aphis* sp.), nematode species and infected pollen. There are also several virus or virus-like diseases of minor importance whose mode of transmission has not been determined. In addition, Rubus stunt is a serious disease associated with a mycoplasma-like organism, while "crumbly fruit" is a serious condition sometimes caused by virus infection and sometimes caused by genetic disorders. There are several reviews of these diseases which may be consulted for further references (Frazier, 1970; Converse, 1966, 1977; Murant, 1974, Jones, 1986a).

8.1 VIRUSES TRANSMITTED BY *AMPHOROPHORA IDAEI* (BÖRN.), *A. RUBI* KALT., *A. AGATHONICA* HOTTES OR OTHER *AMPHOROPHORA* SPECIES

The diseases caused by this group of viruses have been called by various names, but are usually referred to as mosaics. In most cultivars the viruses infect without inducing recognizable leaf symptoms, apart from the diseases caused by infection with combinations of viruses. They are the most widespread of the virus diseases, and usually cause gradual degeneration of stocks rather than strongly damaging effects. The viruses are transmitted in a semi-persistent manner; that is, the aphids acquire them after feeding for an hour or less on an infected plant and transmit them to healthy raspberries without a latent period during a 15 minute feed. *A. agathonica* acquires virus less readily from host cultivars that are resistant to it than from susceptible ones. All the vectors lose their virus charge after feeding on a succession of healthy raspberries for two hours or after one to four days when not feeding. This is because there is no multiplication of virus in the vectors. *A. idaei* and *A. agathonica* are the main vectors for raspberries in Europe and North

America respectively, and *A. rubi* is the main vector for blackberries in Europe.

Raspberry mosaic in this broad sense occurs wherever *Rubus* cultivars are grown, but its presence in Australasia or Chile is due to the introduction of infected stocks, because the important aphid vectors do not occur there. The viruses infect both raspberries and blackberries but are most common in red, black and purple raspberries, even though their incidence and effects tend not to be noticed. The four important viruses in the group are rubus yellow net, black raspberry necrosis, raspberry leaf mottle and raspberry leaf spot, while raspberry yellow spot, cucumber mosaic and thimbleberry ringspot are less important. The presence and identity of these viruses in plants is determined by the differential reactions of graft-inoculated *Rubus* indicator plants. In both Europe and North America, breeding for resistance to the aphid vectors is being given high priority as a means of reducing spread, and experiments show that the resistance is highly effective in restricting virus infection.

8.1.1 Rubus Yellow Net Virus (RYNV)

Of the viruses transmitted by *Amphorophora idaei* or *A. agathonica*, RYNV is the only heat-stable one, that is to say, the virus in infected plants is not lost during heat therapy. It induces symptoms in black raspberries, but only in a few cultivars of red raspberry; black raspberries are therefore used as indicators in graft tests to identify the virus. In black raspberries the initial symptoms are diffuse chlorotic flecks along the veins of subterminal leaflets; these later merge into a vein chlorosis with a diffuse net-like appearance. There is also leaf distortion and a general weakening of the plant and the fruits tend to be crumbly.

In red raspberries a pale net-like vein chlorosis and slight downward cupping of the leaflets may occur in cultivars like "Washington", "Cuthbert" or "Newburgh", but the effect on yield is small. Raspberry tissue infected with RYNV contains small bacilliform particles.

Rubus yellow net virus can be serious in red raspberries when it is present with one or more of the heat-labile viruses of this group. Its combination with black raspberry necrosis virus (BRNV) produces a disease known as mosaic in North America (Stace-Smith, 1955) or veinbanding in Europe (Cadman, 1961). The difference in names probably arises because the vein-banding symptom is characteristic of the disease complex in European but not in North American cultivars. Thus multiple infection in susceptible European cultivars like "Malling Jewel" develops as a pronounced chlorosis of the leaf lamina in areas adjacent to the main leaf veins and produces yellow-green chlorotic veinal bands running out to the leaf margins (Fig. 8.1). Severely affected leaves may be slightly down-curled. Symptoms tend to be reduced or masked during warm sunny weather. By contrast, symptoms in tolerant North American cultivars like "Newburgh" and its derivatives "Malling Promise" and "Malling Exploit" tend to be fainter, more diffuse and irregular, and restricted to leaf margins, while none have been seen on "Glen Clova". Because of this difference, Jones (1982) suggested that vein-banding mosaic disease was a suitable name for the disease caused by the virus complex which includes RYNV. Jones (1986b) later showed that as well as RYNV and BRNV, raspberry leaf mottle virus, sometimes with raspberry leaf spot virus as well, was needed to produce the vein-banding symptom in two British cultivars, and that the most severe symptoms occurred when all four viruses were present together. It is difficult to diagnose the presence of these viruses when they are present together, and

Fig. 8.1. Raspberry leaf with a severe form of veinbanding mosaic disease, caused by combined infection with Rubus yellow net, black raspberry necrosis and raspberry leaf mottle viruses. The pronounced chlorosis occurs in areas adjacent to the main leaf veins and produces veinal bands running out to the leaf margins.

so there must be doubts about the comparability of the earlier North American and European work.

8.1.2 Black Raspberry Necrosis Virus (BRNV)

Black raspberry necrosis virus occurs widely in Europe and North America as a latent infection in many red raspberry cultivars. It was first recognized by the severe symptoms that it induces in black raspberry. In the black raspberry "Munger", the tips of inoculated plants become slightly bent soon after infection and then down-curled and brittle; partially expanded leaves appear wilted and flecks of necrotic tissue appear along the petioles, midribs and secondary veins. These ultimately coalesce to form diffuse areas of blackened necrotic tissue and the cane tip is killed back for two to three inches. This is a shock reaction, because young tissue formed later shows a chronic light-green mottle rather than a necrosis. The virus does not induce symptoms in most blackberries, hybrid berries or red raspberries, but it can cause a moderately severe chlorotic mottle in the blackberry "Himalaya Giant", and a faint transient mottle in the Loganberry and in certain red raspberries, such as "Malling Admiral", "Malling Orion", "Taylor" and "Washington".

All the red raspberry cultivars and *Rubus* species tested are infectable and the ability to suppress symptoms in infected red raspberries is controlled by dominant genes. Hence first-generation hybrids of red and black raspberries react like their red raspberry parent, and second- and third-generation hybrids show mosaic symptoms of varying intensities intermediate between the extremes shown by the two parents: there is therefore a continuous range of symptom expression and the genetic control is complex (Jones and Jennings, 1980).

Black raspberry necrosis virus is most damaging when it is present with other viruses. Its combination with RYNV is described above, and though it can cause a mild form of "bushy dwarf" disease or symptomless decline when present alone in "Lloyd George", the symptoms are accentuated when BRNV is present with raspberry bushy dwarf and other latent viruses. Such mixed infections can cause severe reductions in cane numbers and height and in fruit size, form and structure (Jones, 1979).

8.1.3 Raspberry leaf mottle and raspberry leaf spot viruses (RLMV and RLSV)

Raspberry leaf mottle virus (RLMV) and raspberry leaf spot virus (RLSV) do not induce symptoms in most red raspberry cultivars, but in non-tolerant cultivars they produce a distinct type of disease sometimes referred to as leaf spot mosaic disease (Fig. 8.2). This is characterized by sharply defined chlorotic spots, often angular in shape, which are scattered randomly over the leaf surface and accompanied by crumpling of the leaf lamina. Plants usually die within two to four years of infection. Both viruses induce a tip necrosis in black raspberry and *Rubus henryi* which is indistinguishable from the symptom induced by BRNV, and they can be identified only by the reactions of certain non-tolerant raspberry cultivars to them.

In red raspberries the ability to develop symptoms of RLMV infection is determined by the dominant gene *Lm* and a similar response to RLSV is determined by gene *Ls*. Notable examples of indicator cultivars which carry these genes are "Glen Moy" and "Glen Clova", which carry gene *Ls*, and "Malling Landmark" and "Malling Delight", which carry gene *Lm* (Jones and Jennings, 1980). The viruses have not been transmitted to herbaceous test plants and little is known about their properties, though they are heat labile and their mode of transmission by the insect vector is similar to that of other viruses transmitted by *Amphorophora idaei*.

Recent work (Kurppa and Martin, 1986) suggests that the two viruses

Fig. 8.2. Symptoms of leaf spot mosaic disease caused by raspberry leaf spot virus in a sensitive raspberry genotype carrying gene *Ls*. The sharply defined spots are randomly scattered over the leaf lamina.

occur in North America, but it is not known whether they are recent introductions there, because few reactors to them have been discovered among American cultivars and until recently the indicator cultivars used there to detect latent infections did not distinguish them from BRNV. In Europe, infection of most commercial raspberry stocks is widespread and infected stocks of symptomless cultivars are sometimes a hazard to non-tolerant ones. A typical example was the disease problem which arose when "Glen Clova" was introduced in Britain in 1970. Fortunately, because "Glen Clova" has moderate resistance to the aphid vector, the disease in Scotland was common only in plants growing adjacent to symptomlessly infected plants of such cultivars as "Malling Jewel", which is a good source of both the virus and its vector; but it was more serious in areas where high vector populations occurred.

Cultivars which suppress symptoms of infection by these viruses are capable of acceptable commercial yields, but the effects of the viruses on them are not altogether negligible. The viruses affect growth and reduce yield, largely through effects on fruit size, and their adverse effects are aggravated when several latent viruses are present together, a situation which commonly prevails (Jones, 1979). Control of the viruses is therefore desirable regardless of whether they cause visible diseases.

8.1.4 Raspberry Yellow Spot and Cucumber Mosaic Viruses (RYSV and CMV)

Raspberry yellow spot virus is a serious disease of red raspberries in Poland, and has not been reported elsewhere. In "Malling Exploit", "Malling Promise" and in wild raspberry the symptoms are pale green or yellow spots of regular size and shape, often restricted to areas along the veins but sometimes so numerous that they cover most of the leaf and give a general yellowing, fading to a creamy white. The disease is associated with leaf curling and deformity. In black raspberries the yellow spots usually remain separate along the veins at first and form "oak leaf" patterns, but greatly distorted crinkled leaves are produced in the year after infection. Severe symptoms, including tip necrosis, occur in "Himalaya Giant" blackberry, and in *R. phoenicolasius*. There are no symptoms in *R. henryi* (Basak, 1974).

Although cucumber mosaic virus is widespread in other hosts, it has been reported in raspberry only in Scotland and the U.S.S.R. and is probably of little economic importance.

8.2 VIRUSES TRANSMITTED BY *APHIS IDAEI* V. D. GOOT AND *APHIS RUBICOLA* OESTL.

8.2.1 Raspberry Vein Chlorosis Virus (RVCV)

Raspberry vein chlorosis virus is a heat-stable virus which causes a fine yellow or yellow-green chlorosis of varying intensity in the small veins of raspberry leaves. The yellow-net pattern resembles the one caused by RYNV but is more sharply defined; it is shown particularly well by such cultivars as "Norfolk Giant" and "Malling Delight", where the leaves may also be distorted (Fig. 8.3). Many cultivars are tolerant and the North American cultivars "Latham", "Cuthbert" and "Viking" are immune. Their immunity is inherited in a complex manner (Jennings and Jones, 1986). *R. henryi*, black raspberries and most other *Rubus* species also appear to be immune. The virus is transmitted by the small raspberry aphid, *Aphis idaei*, and, like this vector, it is more prevalent in continental Europe than in Britain. The disease has been introduced into North America, New Zealand and Canada but is not important there, though it has been spread by the propagation of infected stocks. The virus has large bacilliform particles,

Fig. 8.3. Symptoms of vein chlorosis disease in raspberry, caused by raspberry vein chlorosis virus. Sharply-defined chlorosis of varying intensity occurs in the small veins of the leaf.

characteristic of the rhabdovirus group, and like other viruses of this group it probably multiplies in the vector and infects it for life (Murant and Roberts, 1980).

8.2.2 Raspberry Leaf Curl Virus (RLCV)

Raspberry leaf curl was noticed in New York as early as 1880 as a severe disorder of red raspberries that spread from plant to plant. Hence it may be the first raspberry virus disease to have been recognized. It causes considerable reduction in yield and usually kills infected plants in a few seasons. Leaves on both the fruiting and primocanes are curled and slightly yellow; the fruiting laterals are shortened and there may be a proliferation of lateral shoots to produce a rosette. The new canes tend to be numerous and branched and prone to winter kill. Some cultivars are infected symptomlessly.

The disease is now known to be caused by either of two viruses or virus strains, referred to as alpha and beta leaf-curl viruses. The former is limited to red and purple raspberries and the latter infects black raspberries as well. Blackberries are only rarely infected naturally, but "Himalaya Giant" and the indigenous Pacific-coast blackberries are susceptible. Infected plants of *R. henryi* develop a chlorotic curl followed by tip necrosis. Both viruses are restricted to North America, but are not often found west of the Rockies, because the vector, *Aphis rubicola* Oestl, is not common there. The disease is most common in red raspberries in New England and the Rocky Mountain region, and in black raspberries in Michigan and Ohio. Nothing is known about the properties of the virus.

8.3 VIRUSES TRANSMITTED BETWEEN MATURE PLANTS BY POLLEN

8.3.1 Raspberry Bushy Dwarf Virus (RBDV)

Raspberry bushy dwarf virus was so named because it was thought to cause the so-called bushy dwarf disease in "Lloyd George", but recent work indicates that this disease is more likely to be caused by BRNV, with symptoms accentuated by its combination with RBDV or other latent viruses (Jones, 1979). Under Scottish conditions, RBDV appears only to affect the pollen, which though normal in appearance has a reduced capacity to induce fruit-set on either healthy or infected mother plants. This leads to the production of crumbly fruit through failure of up to 43% of the fruits' drupelets to set (Murant *et al.*, 1974; Daubeny *et al.*, 1978). In other environments infection has been associated with "oak leaf" and "line pattern" symptoms (Daubeny *et al.*, 1982). In New Zealand and Canada it is more commonly associated with a yellows disease or line pattern than it is in Britain; for example, the common yellows disease of "Lloyd George" and "Marcy" red raspberries in New Zealand is caused by RBDV infection (Jones and Wood, 1979, Jones *et al.*, 1982). The symptoms of this disease are a yellow vein netting of the lower leaves in spring and a general leaf chlorosis.

Many red raspberry cultivars are immune from the virus immunity being determined by gene *Bu*, while some infectable cultivars like "Norfolk Giant" are highly tolerant or resistant to natural infection and do not produce crumbly fruit when infected. A resistance-breaking strain of the virus has been discovered in raspberries from Russia and West Germany, but, though it can infect nearly all the *Rubus* types tested, it appears to be limited in commerce to Loganberries in Britain. It has probably been unwittingly distributed to other countries in infected stocks of these. Raspberry bushy dwarf virus also infects black raspberries, where it does not produce symptoms but reduces vegetative growth, and some blackberries such as the Boysenberry; however, the isolates obtained from black raspberries differ serologically and in some other respects from those found in red raspberries. A Loganberry degeneration disease in Britain was once associated with infection by RBDV, but similar degeneration has not been reported in Loganberry in North America, and a recent survey in England did not associate any symptoms with Loganberry infection (Barbara *et al.*, 1985).

Raspberry bushy dwarf virus is common wherever susceptible red raspberries are grown. It is transmitted by seed and pollen and it is possible that infection of several important cultivars occurred through the seed (Daubeny *et al.*, 1978). It is transmitted to plants pollinated with infected pollen and consequently spreads rapidly once plants begin to flower. It is readily transmitted by manual inoculation of sap to *Chenopodium quinoa*. The virus is difficult though not impossible to eradicate from infected plants by heat therapy. It has isometric particles of about 30 nm diameter.

8.3.4 Tobacco Streak and Apple Mosaic Viruses (TSV and ApMV)

Tobacco streak virus spreads through pollen and seed but can also spread very slowly in deblossomed *Rubus* plants, suggesting that an additional method of spread exists (Converse, 1986); thrips is a possibility. The virus is not indigenous to Europe but is common in North America and Australasia, mostly in black raspberries and Boysenberries.

A feature of the virus is the diversity of its isolates. The *Rubus* strain (TSV-R) of Converse (1966) occurs in cultivated black raspberries throughout the U.S.A. and is also found in some red raspberry and blackberry cultivars, though the range of cultivar susceptibility is not known. Several strains of the so-called black raspberry latent virus are serologically related to it but are distinct (Jones and Mayo, 1975; Brunt and Stace-Smith, 1976). These occur in many black raspberry cultivars. All strains of the virus are symptomless in their natural *Rubus* hosts, but can be detected by sap transmission to *Chenopodium quinoa* or by grafting to *R. henryi*. They are seed-borne, heat-stable isometric particles which are often irregularly distributed within infected plants, especially in blackberries.

Apple mosaic virus rarely infects *Rubus*, but infections have been found in North America and Germany. The red raspberry "Schonemann" is exceptional in producing bright yellow symptoms. No vector is known, but spread by pollen is suspected.

8.4 VIRUSES TRANSMITTED BY NEMATODES

Spectacular and spontaneous outbreaks of a severe disease of red raspberries in eastern Scotland were reported periodically from the 1920s, notably in the cultivars "Baumforth B" and "Norfolk Giant". It was named "leaf curl" because of its resemblance to the aphid-borne American disease of that name, but the method of spread remained an enigma until, in 1953, after a prolonged and unsuccessful search for an aerial vector, the disease was found to be transmitted through the soil. Subsequently it was found that plants became infected through their roots and that the vectors were free-living nematodes. The disease was shown to be caused by raspberry ringspot virus (RRV), and it was found that there were other nematode-borne viruses, one of which could produce a disease similar to leaf curl.

These viruses are transmitted by relatively immobile nematodes which do not travel for more than a few inches. All the viruses are seed-borne and are disseminated widely through the seeds of many weed and crop species, together with the unwitting use of infected propagation material: this is an easy error because symptoms frequently fail to express themselves until two or three years after infection. Infected seeds also provide a continuing source of virus in the soil, because the viruses do not persist indefinitely in their nematode vectors. Those transmitted by *Longidorus* species (needle nematodes) are retained for only a few weeks, while those transmitted by *Xiphinema* species (dagger nematodes) are retained for many months.

The frequent occurrence of immunity, determined for each virus by a single dominant gene, provides a ready means of control through plant breeding. Resistance-breaking strains have been discovered for at least two of the viruses, but little is known about their geographical distribution. Good weed control and soil treatment by nematicides also provide control; the latter should be done when nematode infestations are high or when raspberries are planted after good nematode hosts such as strawberries or grass crops. However, the treatment has been less effective for the viruses transmitted by *X. diversicaudatum*, probably because the longer persistence of the viruses in this vector means that the few nematodes that survive to recolonize the sterilized area are likely to have retained the ability to transmit the viruses.

The viruses belong to the nepo group, so-called because they are nematode-borne and have polyhedral particles.

8.4.1 Raspberry Ringspot and Tomato Blackring Viruses (RRV and TBRV)

Disease outbreaks caused by these viruses may occupy a few square metres or several hectares, depending on the distributions of the nematode vector and virus in the soil (Fig. 8.4). Both are transmitted by *Longidorus elongatus* (de Man) Thorne & Swanger in Scotland and often occur together, while *L. macrosoma* Hooper, which transmits some strains of RRV, and *L. atte-*

Fig. 8.4. Aerial view of a raspberry plantation in Scotland showing bare areas caused by the death of plants infected with the nematode-borne raspberry ringspot virus.

nuatus Hooper, which transmits some strains of TBRV, are of minor importance in southern Britain. The viruses have not been reported in raspberry outside of Europe. Infected plants of non-tolerant cultivars have characteristic yellow spots and rings ("ringspots") which tend to be conspicuous in spring and autumn but to disappear in summer; some show the characteristic pronounced curling of the leaf, which becomes brittle and snaps under slight pressure. The disease caused by RRV is lethal in non-tolerant cultivars within two years of infection and outbreaks of the disease are seen as bare patches in otherwise vigorous plantations. Plants of tolerant cultivars show only mild symptoms, but their vigour is reduced and they are late to leaf-out in spring. Tomato blackring virus infection is usually less serious, and though affected plants are reduced in vigour they are rarely killed. This is partly because most susceptible cultivars are tolerant of it. The cultivars "Malling Promise" and "Malling Exploit" are tolerant of both viruses, and infected plants show few or no leaf symptoms but reduced vigour and crumbly fruit, owing to effects of the viruses on fertility.

8.4.2 Arabis Mosaic and Strawberry Latent Ringspot Viruses (AMV and SLRV)

Arabis mosaic virus and strawberry latent ringspot virus are locally important in England and parts of Europe, but not in Scotland; they have not been found in raspberry in North America, presumably because their vector is absent there. They are both transmitted by the nematode *Xiphinema diversicaudatum* Micoletzky and are sometimes found in association with each other. Plants infected with AMV show a progressive decline in vigour until they become dwarfed and give little or no yield 3 to 5 years after infection. Leaf symptoms are frequently absent, but there may be small yellow spots on the expanding upper leaves and a conspicuous vein-yellowing or yellow-net on the lower ones. Except for "Malling Jewel", plants infected with SLRV show no symptoms, while those infected with both AMV and SLRV look similar to plants infected with AMV alone. The disease occurs in outbreaks, similar but less spectacular than those caused by RRV and TBRV; AMV has also been found in wild blackberry (*R. fruticosus* agg.), and SLRV in "Himalaya Giant" blackberry.

8.4.3 Tomato Ringspot, Tobacco Ringspot and Cherry Rasp Leaf Viruses (TomRSV TobRSV and CRLV)

Tomato ringspot virus and tobacco ringspot virus are found affecting raspberries, Boysenberries and "Himalaya Giant" blackberries in eastern and western North America, but only rarely Europe, where their vectors *Xiphinema americanum* Cobb and possibly *X. bakeri* are absent. Tomato ringspot virus is widespread in North America; it induces ringspot symptoms on the leaves and varying degrees of dwarfing depending on the host cultivar. The most serious effect on raspberries is drupelet abortion and the production of crumbly fruit, but this varies considerably with cultivar; it is severe in "Lloyd George", "Avon" and "Sumner", but generally less so on most other cultivars (Daubeny *et al.*, 1975). It affects the fertility of both pollen and ovules. In "Himalaya Giant" blackberries the leaf symptoms are chlorotic spots and blotches, with veinal chlorosis and oak-leaf patterns.

Tobacco ringspot virus is not common in *Rubus*. Knowledge of its occurrence is limited to occasional reports in blackberries near tobacco fields in North Carolina and another report from British Columbia. Similarly, CRLV has occasionally been found in raspberries only in Canada. Both are transmitted by *X. americanum*.

8.4.4 Cherry Leaf Roll Virus (CLRV)

Cherry leaf roll virus causes a serious disease of raspberry only in New Zealand. It is associated with a disease of the raspberries "Lloyd George", "Marcy" and "Taylor", where the symptoms are stunted fruiting canes and distorted leaves, sometimes with a severe chlorotic mottle and ringspots. It has been found in "Himalaya Giant" blackberry in Britain, but has not so far been found infecting *Rubus* plants in North America. Its mode of transmission is not known; it is placed in the nepo group but it may be pollen-transmitted (Jones, 1985).

8.5 VIRUSES THAT CAUSE "CRUMBLY FRUIT"

A condition known as "crumbly fruit" or "fruit shrivel" is caused by failure of some drupelets to set, resulting in misshapen fruit. The drupelets may be greatly enlarged if their number is greatly reduced or, if small reductions are involved, they cohere imperfectly so that the fruit readily crumbles when picked. The condition may be caused by virus, or by genetic disorder. Among the viruses causing the condition are RBDV, TomRSV and TBRV and, of these, RBDV is the most important in Europe and both RBDV and TomRSV are important in America. Infection of eastern North American blackberries with a graft-transmissible agent tentatively designated blackberry sterility virus can reduce fruit-set to a very few drupelets per fruit, and was probably the main cause of the disappearance of the important blackberry "Eldorado". It also causes leaf malformation in infected plants of the blackberry "Early Harvest" and can infect black raspberries.

8.6 VIRUS AND VIRUS-LIKE DISEASES OF LOCAL OR MINOR IMPORTANCE

Several other viruses and graft-transmissible diseases have been reported but they are either only occasional, or are serious in particular areas or particular cultivars. Their overall importance can be regarded as minor, but some of them are of major importance locally. Notable examples are the calico diseases of North America; one of these was associated with the raspberry cultivar "Puyallup" where it caused a brilliant-yellow vein clearing on the leaves followed by dwarfing. Another is Loganberry calico which also occurs in blackberries, but although it causes yellow spots or chlorotic oak-leaf patterns especially on the floricane leaves, it does not markedly affect growth. "Chehalem" and "Marion" are commonly affected. The diseases are graft transmissible with a long latent period (Converse, 1984).

Some other examples are tobacco rattle virus, Brazos mottle, necrotic fern-leaf mosaic, Olallie disease, Bedford giant mottle, bramble yellow mosaic and black raspberry streak. Most of these have been graft transmitted, but details of their natural mode of transmission are not understood in every case. There are other instances where viruses are found in wild plants and not in domesticated forms; the aphid-borne thimbleberry ringspot virus for example infects wild thimbleberries (*R. parviflorus*) in British Columbia, and wineberry latent virus has been found in *R. phoenicolasius*.

8.7 *RUBUS* STUNT AND MYCOPLASMA-LIKE DISEASES TRANSMITTED BY THE LEAFHOPPER *MACROPSIS FUSCULA* ZETT.

The striking symptoms of *Rubus* stunt have made it one of the historic diseases of raspberry, because it was noticed and referred to by de Vries as early as 1896. It is an important disease of raspberries in the Soviet Union and eastern Europe, but is rare in Western Europe and does not occur in North America. It affects blackberries, red raspberries and hybrid berries, in all of which the typical symptoms are a stunted and witches' broom type of growth with numerous thin canes, excessive branching and a phylloid type of irregularity in the flowers. The disease is associated with a mycoplasma-like agent and is transmitted by the leafhopper *Macropsis fuscula* Zett., in which it persists for life (Murant and Roberts, 1971).

Mycoplasma-like bodies have also been associated with a witches' broom disease of the Thornless Evergreen blackberry in the Netherlands (Dijkstra, 1973) and in "Munger" black raspberry (Converse *et al.*, 1982).

8.8 GENETIC DISORDERS

Genetic disorders caused by somatic mutations are important when they affect the fruit or any part of the fruiting laterals without inducing an easily recognizable effect on primocane growth. This is because they can occur during propagation and many cycles of vegetative multiplication can take place before their presence is suspected. "Outbreaks" may then occur from the unwitting vegetative propagation of affected plants. There are no tests to detect the disorders apart from inducing the plants to fruit, and this precaution should be introduced into the procedure for producing certified stocks.

The best known genetic disorders are those that reduce fertility and cause crumbly fruit in much the same way as viruses can cause it. In Scotland crumbly fruit has occurred in "Norfolk Giant" and "Malling Promise", apparently because of a chromosomal mutation. In "Norfolk Giant" it occurred in virus-tested mother plants and all the cane nurseries established from them were consequently found to contain a high proportion of valueless plants (Murant *et al.*, 1973). The disorder has also occurred in "Malling Jewel" and "Latham", but in these instances it was associated with mutation of the dominant allele at a heterozygous gene locus, causing plants to become homozygous for a deleterious recessive gene (Jennings, 1967b). The cultivar "Sumner" is particularly prone to crumbly fruit and it has been suggested that crumbliness in this instance is caused by a mutation giving homozygosity for two recessive gene pairs whose effects are to retard embryo sac development and reduce the production of fertile pollen (Daubeny *et al.*, 1967). A sterility disorder of "Darrow" blackberry is also thought to be a genetic disorder (Converse, 1986).

Although genetic disorders are mainly associated with crumbly fruit, other mutations that affect fruiting canes and cannot be identified in primocanes have been identified as potential problems during propagation. These include "whorled receptacle" and "catkin" in "Malling Jewel", and "lateral-leaf crinkle" in "Glen Clova" (Jennings, 1977).

8.9 CERTIFICATION SCHEMES

The use of certified healthy planting material is the most effective means of controlling all virus and genetic diseases, provided that healthy "mother" stocks are available to initiate propagation. Efficient schemes for the propagation and certification of virus-tested stocks now operate in many countries where *Rubus* fruits are commercially important. The schemes are based on the identification of apparently virus-free plants and the subsequent propagation of stocks derived from them under conditions where reinfection is kept to a minimum. In most schemes the plants are considered to be virus-free if they give negative results in graft tests to *R. henryi* and/or black raspberry indicator plants, and in sap inoculation tests to *Chenopodium quinoa*.

Where stocks were uniformly infected with a heat-labile virus (BRNV, RLMV or RLSV), or by Rubus stunt, it has been possible to rid them of infection by appropriate heat treatment; and where the virus is not fully systemic, as with tobacco streak virus, it has been possible to obtain healthy stocks by extensive propagation and indexing. Meristem culture has been used alone or in combination with heat therapy to free plants of non-heat-labile viruses (Achmet *et al.*, 1982; Baumann, 1982; Sobczykiewicz, 1986).

Procedures in Scotland were modified from 1965 onwards to safeguard against the possibility of propagating stocks in which mutations had occurred to genetic disorders such as crumbly fruit. The procedure is simply to fruit the plants used for root production, contrasting with earlier practice, and to reject any that show a genetic disorder or fail to set perfect fruit. The procedure does not detect mutant tissue present in a portion of the root system and not in the tested cane, so there is always a risk that a small proportion of affected plants may be propagated. Usually the fruiting tests are done concurrently with propagation from the roots, and all the plants derived from substandard stocks are discarded (Murant *et al.*, 1973).

In most schemes, several thousand plants derived from the tested mother plants are distributed annually to appropriate propagating agencies for growing in cane nurseries under the conditions required by the certification schemes. These conditions normally include minimum standards of isolation from other *Rubus* plants, including wild sources, tests for nematodes in the soil and appropriate control of aphid virus vectors. All fruiting canes are removed from the nursery to prevent mixing with self-sown seedlings, and to prevent the spread of pollen-borne viruses. This presents problems with primocane-fruiting cultivars whose flowering tips should also be removed. Paths for official inspectors to examine the crop for certification purposes are also required. Stocks are eligible for certification for a limited number of seasons after propagation from the foundation mother plants.

9 Diseases Caused by Pests

Many pests are found on raspberries and blackberries, and the removal of DDT as a means of control has caused many of them to become prevalent after a period when they were largely controlled by this insecticide. Only the more important ones are described here. Additional information and references are given by Massee (1954), Cram and Neilson (1976) and Pears and Davidson (1956).

9.1 RASPBERRY BEETLES (*BYTURUS* SPECIES)

In Europe the important species of beetle is *B. tomentosus* Degeer (Fig. 9.1). Its importance was recognized as early as 1827, when one author wrote of the stalks of raspberry blossoms being eaten through by a minute animal to such an extent that it proved fatal to the whole crop. The beetle is about 5 mm long, light brown when it first emerges from the soil and grey-brown or grey later. In Britain emergence begins as temperatures rise in May and continues until July. The beetles feed first on the growing points of primocanes of raspberries, blackberries and hybrid berries, but fly actively to the unopened flower buds and congregate in the flowers as soon as they open. This feeding either destroys the flower buds or leads to malformation in flower and fruit development. The female lays eggs in the flowers in June and July and larvae hatch within 10 days, usually when the fruit is at the green or pink stage. They browse first on the outside of the green fruit and then burrow into the receptacle and continue feeding inside the fruit. This can cause failure of drupelet development, leaving groups of drupelets as hard areas in misshapen fruit; but the main problem is the mere presence of larvae contaminating the fruit. This is enough to condemn any sample, whether for use fresh or for processing. Any mature larvae which are not gathered with the fruit drop to the soil and form a cocoon.

Beetles can be controlled by spraying at the pre-blossom stage, which avoids killing honeybees and is easy to do because primocane growth is not too tall at this time. Unfortunately some beetles emerge after flowering in Scotland and sprays applied between 80% petal fall and the first pink fruit stage are more effective (Taylor, 1971; Taylor and Gordon, 1975), though the application of chemicals at this late stage is becoming unacceptable to

Fig. 9.1. An adult raspberry beetle (*Byturus tomentosus*) feeding on an unopened raspberry flower bud.

consumers. The recent discovery of beetle larvae parasitized by *Tetrastichus halidayi* (Graham) suggests the prospect of biological control (Gerard, 1985).

Other species of fruit worm occur in North America. These are *Byturus bakeri* Barber, the western raspberry fruitworm and *B. rubi* Barber, the eastern raspberry fruitworm. Schaefers *et al.* (1978) noticed that early ripening cultivars tended to be more susceptible to the eastern raspberry fruitworm than late ones, and that there was no attack on primocane-fruiting cultivars during the autumn-cropping period. This suggests that elimination of the summer crop could give complete control.

Strong resistance to *B. tomentosus* has been found in *R. phoenicolasius*, *R. coreanus*, *R. crataegifolius* and *R. occidentalis* (Briggs *et al.*, 1982; Jennings *et al.*, 1977, 1978). For *B. rubi* cultivar differences in adult preference but not in acceptability for oviposition and larval attack have been reported (Schaefers *et al.*, 1978).

9.2 CLAY-COLOURED WEEVIL (*OTIORHYNCHUS SINGULARIS* L.) AND OTHER WEEVILS

Clay-coloured weevils are frequently a serious pest of raspberries in Europe and are occasionally a problem in North America. The weevil is a brown wingless insect speckled with light markings and frequently indistinguishable from soil adhering to it (Fig. 9.2). It emerges from the soil in March or April and feeds on the petioles of leaves on the fruiting laterals, causing them to break or wilt, and damages the developing flower buds. It also checks the growth of the primocanes. The pest is difficult to find because it leaves the soil only at night and hides in soil at the base of the cane by day. The larvae feed on the roots without causing any apparent damage.

A pest that occasionally causes similar damage to raspberries in Britain is the double dart moth (*Graphiphora augur* F.). This moth is also nocturnal. It over-winters as hibernating larvae which emerge in the spring and feed on developing buds and leaves; the latter are sometimes left as a skeleton of veins.

Nemocestes incomptus (Horn) occurs in North America and is similar in appearance and habits to the clay-coloured weevil. This species also attacks raspberry roots, concentrating on lateral roots and usually leaving the main roots, which appear stripped. However, the black vine weevil, *Otiorhynchus sulcatus* Fabricius is the most damaging of the root weevils. Although the pest is found in Europe it is more common in North America, where it is the larvae that cause serious damage to raspberry roots and crowns; the adults rarely do more than chew notches in the leaf margins in spring. Plants whose roots have been damaged are easily lifted from the soil; their leaves wilt and

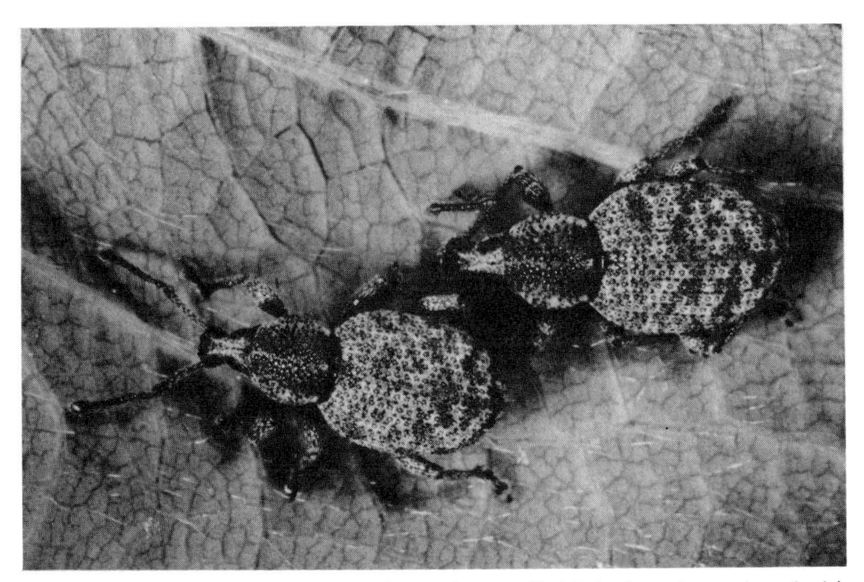

Fig. 9.2. Adults of the clay-coloured weevil (*Otiorhynchus singularis*). (Crown Copyright 1987. Reproduced by kind permission of ADAS, MAFF).

their fruit remain small. "Leo" and "Glen Prosen" show some resistance to this weevil (Cram and Daubeny, 1982).

9.3 RASPBERRY MOTH (*LAMPRONIA RUBIELLA* BJERKANDER) AND BUD MOTHS (*HETEROCROSSA ADREPTELLA* WALKER AND *EUTORNA PAULOCOSMA* MEYRICK)

Raspberry moth, sometimes referred to as raspberry borer, is widespread on raspberries and Loganberries in Europe, but it is a relatively new pest in North America, where it first appeared in the 1930s. It has been known to attack over 90% of raspberry buds in Scotland (Hill, 1952). The adult moth has shiney purple-brown wings marked with two very conspicuous yellow spots and several smaller ones. Overwintered larvae emerge in spring from rubbish or soil at the cane bases. They crawl up the canes and burrow through the closely packed leaves of the laterals, especially those near the cane bases, and cause them to wilt and die towards their tips. When mature, a larva scoops out a cavity in the pith of a primocane just below a bud, or in the lateral itself, and pupates; it may produce a distinct hole at the base of the bud, leaf or shoot where it entered. Moths emerge in May and June and the females lay eggs in open flowers. Larvae emerge from these in seven to ten days and burrow into the receptacle. They feed there until the fruit begins to ripen, but cause little or no visible damage. They are normally controlled by the sprays applied to control raspberry beetles and by good plantation hygiene.

The bud moths, *Heterocrossa adreptella* on red raspberries and *Eutorna paulocosma* on blackberries, are very destructive of fruit buds in New Zealand but they have not been reported elsewhere. Black raspberries and *R. glaucus* are resistant.

9.4 RASPBERRY CROWN BORER (*PENNISETIA MARGINATA* HARRIS), CANE BORER (*OBEREA BIMACULATA* OLIVIER) AND RED-NECKED CANE BORER (*AGRILUS RIFICOLLIS* FABRICIUS)

The raspberry crown borer, sometimes referred to as a raspberry root borer, is a serious pest of all *Rubus* fruits in North America but does not occur in Europe. Its seasonal development does not follow a defined pattern and most of its stages can be found at any time during the year. The moths are clear-winged with black and yellow bands on the abdomen, resembling a wasp but darker and with less yellow and more black in their coloration. Larvae hatch from September to November and attack cane bases and fleshy roots, where they eat through the rind but not the wood, and produce small raised galleries for hibernation. They remain in these for variable times and emerge during a mild spell in winter or spring to attack young canes near ground level. They eat the bark and girdle the canes and then bore into the cane centres, where they form larger galleries several inches long. They usually reach maturity by the end of the second summer and pupate inside the cane. Infested canes are often hollowed out and debilitated; they wilt and are easily broken. The pest is controlled by sprays directed at the cane bases in October and by burning all wilted canes (Breakey, 1963).

The raspberry cane borer occurs in eastern parts of North America and Europe. The adult is a slender black beetle with prominant antennae and usually two black dots on a yellow prothorax. The female feeds on the cane above and below its puncture for egg laying and girdles it, causing the shoot tip to wilt and die. The larvae hatch and bore downwards into the cane and pass their first winter just below the girdling point. They continue to move downwards in the second year and pass the second winter near the ground.

The red-necked cane borer occurs widely in eastern North America, particularly on blackberries. It is seldom serious. Swellings of from one to three inches long, referred to as gouty galls, occur on the canes and cause weakening or death.

9.5 LEAF ROLLERS (*CLENOPSEUSTIS OBLIQUANA* WALKER, *PLANOTARTRIX EXCESSANA* WALKER, *EPIPHYAS POSTRITTANA* WALKER) AND ORANGE TORTIX (*ARGYROTAENIA CITRANA* FERNALD)

Leaf rollers are a major pest in New Zealand, as is the orange tortix in North America. The latter is a pest of Californian orange groves and was once only a glasshouse pest of raspberries. It began to appear on field-grown raspberries and Loganberries in Washington State in the early 1930s and was causing major concern by 1939. For each roller species, larvae feed on cane buds during the winter and then move to the terminal growth in summer; here they produce a white web which causes the leaves to stick together or to the fruit and so form a shelter for feeding and egg laying: hence the larvae's description as "leaf rollers". The presence of larvae or eggs as a contaminant of fruit samples is the main problem, because the tolerance of consumers is near to zero. Black raspberries are resistant.

9.6 RASPBERRY CANE MIDGE (*RESSELIELLA THEOBALDI* BARNES) AND LOGANBERRY CANE FLY (*PEGOMYA RUBIVORA* COQ.)

Raspberry cane midge was first described in south-east England in 1920, but it did not attain any economic importance until the late 1940s and remained unknown in Scotland until 1972. It is now a serious pest throughout Europe (Woodford and Gordon, 1978), largely because of its role in midge blight (see p. 91).

The first generation of adult midges emerges from the soil in May and June in Scotland, usually some two weeks later than in southern England. It lays eggs mostly in the splits that occur from internal growth stresses of the primocanes (see p. 145). Hence its prevalence is partly dependent upon the synchronous emergence of the over-wintering midges and the onset of splitting in the rind of the canes, and partly on the proneness of the cultivar's rind to develop frequent splits. Cultivars whose canes show few splits or split late tend to escape. Larvae are seen within two weeks; they are translucent at first and become pink or orange as they mature (Fig. 9.3). They feed under the rind until they are about 3 mm long and then fall to the ground, burrow below the surface and pupate. The pupae spend only two to three weeks in their cocoons and the second generation normally hatches in mid-summer, by which time there are ample cane splits for egg laying and freshly exposed periderm for feeding. There may be a third generation in August–September, but this varies in size and timing, depending on the weather during the second generation (Pitcher, 1952). First-generation larvae cause canker-like wounds which only slightly affect yield the following year. The lesions formed by second-generation larvae are smaller, but their numbers and cumulative effects are greater.

Fig. 9.3. Larvae of the raspberry cane midge (*Resseliella theobaldi*) feeding on tissues of a raspberry cane exposed by natural peeling of its rind.

The Loganberry cane fly, alternatively known as cane maggot, is occasionally prevalent on red raspberries in northern U.S.A. and southern Canada, especially British Columbia, and is occasionally serious on Tayberry and blackberries (Gordon and McKinlay, 1986) in Britain. The adult resembles a small house fly. It emerges in spring and lays eggs in the growing tips of raspberries, blackberries or hybridberries; the resultant larvae burrow into the cane and mine a chamber around the periphery. This causes the canes to wilt and sometimes to break off as cleanly as if cut by a knife. Multi-branched canes may result. Pupae are found near the site of attack.

9.7 RASPBERRY LEAF AND BUD MITE (*PHYLLOCOPTES GRACILIS* NAL.)

Raspberry leaf and bud mite is a minute insect invisible to the naked eye. It over-winters in colonies in bud scales or in petiole scars, and migrates to the undersurfaces of emerging new leaves of fruiting canes in April, where it lays eggs and grazes amongst the downy hairs. This causes patches or blotches that resemble virus or mildew symptoms. No leaf or bud galls are formed, though dense hairs may be induced to form on the undersurface. When infestations are severe the canes may be stunted and the fruit malformed, or the fruit may ripen unevenly because of earlier infestation of the flower buds. This is why the pest is referred to as the dryberry mite in America.

The mite is widespread on raspberries and blackberries in Europe, the U.S.S.R. and North America, but it tends to attain large populations only in very sheltered sites, especially in the shelter of a garden. In these situations the populations can attain many times the size of those in unsheltered sites (Gordon and Taylor, 1976). This probably explains why mites are regarded as a persistent pest in France but not in Scotland, where hot dry summers conducive to their multiplication are rare. They seriously affect Loganberry production in western Washington and can cause a severe leaf-blotch disorder of Tayberries (Jones *et al.*, 1984). Gordon and Taylor (1977) improved fruit yield and quality by a spray treatment of the primocanes.

9.8 REDBERRY DISEASE OR BLACKBERRY MITE (*ACALITUS ESSIGI* HASSAN)

This is a white translucent mite which infests blackberries and hybridberries in southern England, Europe and North America. It has similar habits to *Phyllocoptes gracilis*, except that it moves from the leaves to the flowers at flowering time. It rarely damages the flowers, but feeds on drupelets, mostly those near the base of the fruit. The resultant fruit are malformed and uneven ripening and show a condition known in North America as redberry disease.

9.9 TWO-SPOTTED SPIDER MITE (*TETRANYCHUS URTICAE* KOCH)

This mite is often serious in North America and Australasia, but less so in Europe. It is light tan or greenish in colour with a dark spot on each side of its body. It is most common on red raspberries, where it can increase rapidly during prolonged hot weather and cause premature leaf browning and defoliation. This reduces yield in the following year and may also increase the susceptibility of the bud tissues to frost (Doughty *et al.*, 1972). Varieties differ in susceptibility (Labanowska, 1978). A method of biological control is to introduce the predator *Phytoseiulus persimilis*.

9.10 RED SPIDER MITES (*PANONYCHUS CAGLEI* MELLOTT AND *P. ULMI* KOCH)

These two species of red spider are morphologically very similar; *P. caglei* occurs in North America and *P. ulmi* occurs in Europe. They are frequent glasshouse pests, but rarely important in the field except in the more southerly parts of America.

9.11 LARGE RASPBERRY AND BLACKBERRY APHIDS

Joshua Major, one of the earliest students of raspberry aphids, described the raspberry aphid in 1829 as "a large green species considerably larger than any aphid yet described". The species was named *Amphorophora rubi* Kalt. in 1906, when it was thought that all the forms found on raspberries and blackberries in Europe and North America belonged to one species. Subsequent taxonomic studies showed that the indigenous North American form was distinct, and it was designated *A. agathonica* Hottes (Kennedy *et al.*, 1962). Its chromosome complement is $2N$ (♀) = 14. In Europe, aphids colonizing raspberries were found to have a chromosome complement of $2N$ (♀) = 18, and those colonizing blackberries were found to have a different complement of $2N$ (♀) = 20. Biometric studies showed that the two forms could also be separated morphologically and should be regarded as separate

Fig. 9.4. A dense colony of *Amphorophora idaei* on the stem and leaves of a susceptible raspberry.

species. This supported an earlier conclusion by Börner based on host–plant transfers, and so the name *A. idaei* Börn. was adopted for the raspberry species and *A. rubi* Kalt. was retained for the blackberry species (Blackman *et al.*, 1977).

In Scotland, eggs of *A. idaei* are laid by apterous oviparae near the bases of raspberry canes in October. They hatch in spring and populations of the aphid first build up asexually on the fruiting canes. The populations reach a peak during July and August on the primocanes but a rapid decline occurs during September and sexual forms occur by the end of this month (Dickson, 1979).

9.12 RESISTANCE TO LARGE RASPBERRY APHIDS

Since *A. idaei* and *A. agathonica* are the vectors of four important viruses of the raspberry, plant breeders have been interested in breeding for resistance to them for many years. The first studies were concerned with resistance to *A. agathonica*, but extensive work was later done with *A. idaei*. It was found that strong resistance to the latter was controlled by major genes, which in many instances confer a near immunity in which only a few aphids occur late in the season on lower leaves. Weaker forms of resistance are conferred by minor genes; in these instances relatively low numbers of aphids occur and

Table 9.1. Resistance (R) or susceptibility (S) to *A. idaei* associated with some major genes.

Gene	Aphid strain			
	1	2	3	4
A_1	R	S	R	S
A_2	S	R	S	S
A_1A_3	R	R	R	S
A_3A_4	S	R	S	S
A_5	R	S	S	S
A_6	R	S	S	S
A_7	R	S	S	S
A_{10}	R	R	R	R

are distributed predominantly on mature or senescing leaves rather than on young leaves, where they are the more numerous on susceptible cultivars (Briggs, 1965).

There are four known races of *A. idaei*. Some major genes confer resistance to the four strains, others only to some of them and some are complementary in their action (Table 9.1 and Appendix 2). Minor-gene controlled resistance is not influenced by differences in aphid strain. The aphid strains can be differentiated by sequential testing for ability to colonize two hosts carrying genes A_1 and A_2A_5 respectively: failure to colonize the A_1 host identifies them as strains 1 or 3 and success identifies them as strains 2 or 4; subsequent failure to colonize the A_2A_5 host identifies them as strains 1 or 2 and success identifies them as strains 3 or 4. Briggs thought that there must be barriers to the development of populations of strains 2 and 4 in the field, because they are rarely found as abundantly as the other strains. He also concluded from cross-breeding experiments that the differences between the four strains were conferred as follows by two genes, designated C_1 and C_2:

Strain 1	$c_1c_1C_2C_2$ or $c_1c_1C_2c_2$
Strain 2	$C_1c_1C_2C_2$, $C_1c_1C_2c_2$, $C_1C_1C_2C_2$ or $C_1C_1C_2c_2$
Strain 3	$c_1c_1c_2c_2$
Strain 4	$C_1c_1c_2c_2$ or $C_1C_1c_2c_2$

Like resistance to *A. idaei*, strong resistance to *A. agathonica* is determined by a single dominant gene designated Ag_1, and weak resistance is conferred by minor genes (Daubeny, 1966, 1980). Gene Ag_1 occurs in several *R. idaeus* cultivars of which "Lloyd George" is the source used most in breeding. Two complementary gene pairs, Ag_2 and Ag_3, which confer strong resistance, have also been found in a wild population of *R. strigosus*

(Daubeny and Stary, 1982). Aphids fed on "Canby", a resistant derivative of "Lloyd George", were found to have low levels of sugars and nitrogenous compounds in their ingestate, suggesting that nutrients were less readily available to them in resistant hosts (Kennedy and Schaefers, 1975). No strains of *A. agathonica* capable of overcoming major-gene-controlled resistance have been found.

9.13 SMALL (LEAF-CURLING) RASPBERRY APHIDS (*APHIS* SP.)

Aphis idaei (v.d. Goot) is a small sedentary aphid which occurs in Europe and forms dense colonies on fruiting canes and primocanes in spring and early summer. It feeds on the undersurfaces of young leaves, causing a characteristic leaf curl, stunting and twisting of the shoot tips, which become a sticky mass of honeydew. It produces a high percentage of winged forms in early summer which later migrate to the young canes and leave only solitary apterae on the leaves. Only apterous viviparous females are produced at first, but sexual forms develop in the autumn (Hill, 1953). *A. rubicola* Oestl. is the important North American species.

Little work has been done to breed for resistance to *Aphis* vectors, partly because they are so difficult to handle and culture, and partly because no strong resistance has been found. Partial resistance has been recorded to *A. rubicola* (Kennedy *et al.*, 1973; Brodel and Schaefers, 1980), and to *A. idaei* (Baumeister, 1961; Rautapää, 1967), but neither has been evaluated for its effect on virus spread. Immunity from the virus transmitted by *A. idaei* is a better prospect for control by plant breeding (Jennings and Jones, 1986; see p. 115).

Several related species, *A. rubifolii* Thomas and *A. spiraecola* Patch also occur on raspberries in North America and *A. ruborum* Börner occurs on raspberries in both Europe and South America. None of these is known to transmit virus.

9.14 APHIDS OF MINOR IMPORTANCE

Several aphids are of minor importance on raspberries or blackberries. A common one is *Macrosiphum euphorbiae* Thomas. This is common throughout Europe and North America but its populations are usually small. It over-winters on herbaceous hosts and migrates mainly to the fruiting laterals of raspberries and blackberries (Hill, 1953). Few are found on the canes after August.

Aulacorthum soloni Kalt. colonizes raspberries in Holland but has seldom been found in Scotland except in glasshouses. Large colonies have been found on certain blackberries in New Zealand. Similarly, *Myzus ornatus* Laing has been found on raspberries in Holland, New Zealand and East Africa but not in Scotland (Dickson, 1979). Both *M. euphorbiae* and *A. soloni* can transmit some of the raspberry viruses.

9.15 LEAFHOPPERS

The leafhopper *Macropsis fuscula* Zett. is widely distributed in certain parts of Europe, Russia and North America and is important as a vector of *Rubus* stunt disease (p. 122), though the disease does not occur in North America. The larvae hatch in mid-summer and feed mostly on flower buds and fruit, sucking up large amounts of plant sap and excreting honeydew which makes the fruit unfit for sale.

Two other leafhoppers, the rose leafhopper (*Edwardsiana rosae* L.) and the bramble leafhopper (*Ribautiana tenerrima* H.-S.), cause serious injury in North America by sucking the sap from raspberry or blackberry leaves, which become speckled with white spots or mottled areas and then curl if the leafhoppers are abundant. They reduce fruit size and cover the fruit with honeydew. The adults are pale and yellow-green with red eyes.

9.16 RASPBERRY SAWFLY (*MONOPHADNOIDES GENICULATUS* Htg.) AND BLACKBERRY LEAF MINER (*METALLUS RUBI* FORBES)

The raspberry sawfly is a spiney pale-green larvae which feeds between the veins of raspberry leaves or eats out irregular holes, often leaving the leaf as a skeleton and causing considerable yield loss when it is present in large numbers. The adult is a black four-winged fly with yellow and red markings. Its larvae aestivate in cocoons which they construct in the ground. The blackberry leaf miner is really a very small sawfly. It is common in North America but not in Europe and its larvae can cause serious injury to blackberry leaves.

9.17 COMMON GREEN CAPSID (*LYGOCORIS PABULINUS* L.)

The common green capsid sometimes attacks cane fruits. Larvae hatch in spring and puncture the young leaves and buds; the perforations then turn brown, and leave large holes in the leaves as they expand. Abundant branching may occur if the apical bud of the cane is damaged. The larvae become adults and disperse to various hosts or to young canes in summer and produce a second generation. The damage caused to raspberries can be particularly severe when the insect passes its entire life-cycle on the crop (Hill, 1952).

9.18 BLACKBERRY GALL MAKERS (*DIASTROPHUS TURGIDUS* BASSETT, *D. NEBULOSIS* (O.-S.) AND *D. CUSCUTAEFORMIS* O.-S.)

These insects cause medium to large polythalamous galls on blackberry stems in North America. The galls are first green and later red or brown, and, though conspicuous, they are apparently harmless.

10 Diseases Caused by Nematodes

Many species of plant parasitic nematodes are found in *Rubus* plantations, but only a few are associated with economic damage. For these species, the severity of the damage that they cause is proportional to their population density in the soil. Damage caused by their extensive feeding usually produces symptoms similar to those caused by any other root injury which impedes nutrient and water uptake. Stunting of growth is the obvious symptom, but various deficiency symptoms such as leaf yellowing may occur if nutrient uptake is affected. Root stunting also occurs, sometimes in association with swellings or galls at the root tips if nematodes that feed at the root tips are present, and sometimes in association with necrotic lesions if the nematodes enter the roots. Newly planted raspberries are particularly vulnerable and the damage often results in poor establishment.

10.1 *PRATYLENCHUS* SPECIES

Six species of *Pratylenchus* nematodes have been associated with raspberry. *P. penetrans* (Cobb) Fil. & S. Stek. is the most pathogenic of them and causes stunting and sometimes cane death (McElroy, 1977; Trudgill, 1983). *P. vulvusoof* and *P. crenatus* are of minor importance. Typically a patch of stunted growth appears in the year after planting and continues to be unproductive in subsequent years.

Juvenile and adult *Pratylenchus* nematodes penetrate the root by puncturing the wall of an epidermal cell with their mouth stylet. They then invade the cortex and kill the tissues by their feeding. The feeding areas develop into characteristic necrotic lesions which have earned *Pratylenchus* species the common name of root lesion nematodes. The lesions occur mainly on young feeder roots and appear first as brownish spots which enlarge and spread along the root axis. Individual lesions coalesce and eventually girdle and kill the distil part of the root. Secondary pathogens frequently aggravate the damage. The life-cycle of *P. penetrans* takes from 4 to 12 weeks, depending on temperature. Eggs are deposited by gravid females within the roots or in the soil nearby. The first juvenile moult occurs within the egg, and it is the active second-stage juveniles which hatch and extend the lesion, or they may leave the root and invade other roots to establish new infestation sites. The juveniles moult a further three times before they reach the adult stage, when they are 0.4–0.8 mm long.

10.2 *XIPHINEMA* SPECIES

Three species of *Xiphinema* occur on raspberries. These are *X. bakeri* Williams, *X. diversicaudatum* (Micol.) Thorne and *X. americanum* Cobb, though recent studies suggest that the latter is a complex of several species (Lamberti & Bleve-Zacheo, 1979). They are relatively large ectoparasitic nematodes, ranging in length from 2 mm for adults of *X. bakeri* and *X. americanum* to 5 mm for *X. diversicaudatum*. The long, hollow feeding stylet in the head gives them the common name of dagger nematodes, and enables them to feed deep into the tips of young roots (Griffith and Robertson, 1984) (Fig. 10.1). The saliva which is injected stops root growth and causes small galls to form. In British Columbia *X. bakeri* is widespread, particularly in loam soils, and even a relatively small population of 200 nematodes per litre of soil can reduce root and cane growth. *X. americanum* also occurs in North America but causes little direct damage except as a virus vector. *X. diversicaudatum* is also a virus vector, but its direct feeding damage can lead to a decline in growth of the raspberry. It is widespread in many parts of Britain and Europe but does not occur in the main raspberry growing area of Scotland (Taylor and Brown, 1976) and is rare in North America.

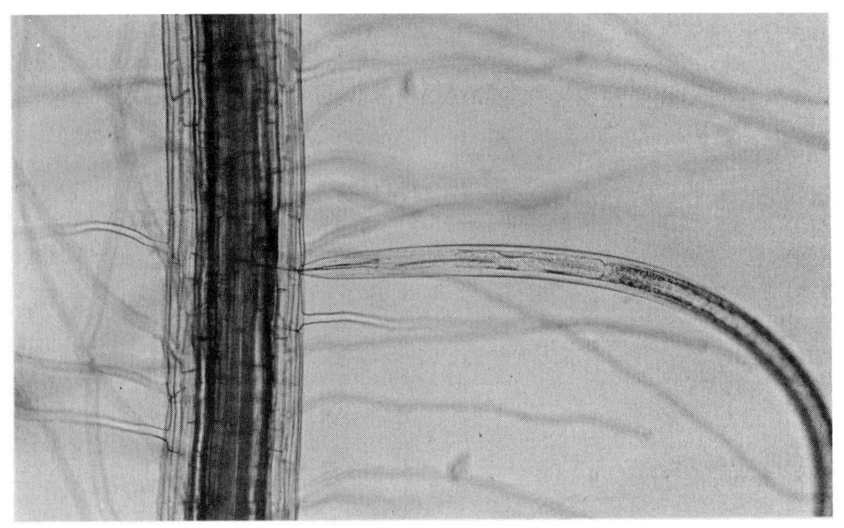

Fig. 10.1. A dagger nematode, *Xiphinema diversicaudatum*, feeding on the root of a susceptible host.

Xiphinema species usually have wide host ranges and long life. *X. diversicaudatum* and *X. americanum*, for example, require a year or more to complete each generation. Egg laying usually occurs when the host plants are producing flushes of new roots in spring. Juveniles hatch from the eggs soon after they are laid and do not develop further unless they have access to a suitable host. Both juveniles and adults can survive for long periods without feeding.

10.3 *LONGIDORUS* SPECIES

In many ways *Longidorus* nematodes are similar to *Xiphinema* nematodes. *Longidorus* species are known as needle nematodes and are similar to *Xiphinema* species in respect of their size range, hypodermic-like feeding stylet, wide host range and long life-cycle. *L. elongatus* (de Man) Thorne & Swanger and *L. macrosoma* Hooper are important as virus vectors in northern European raspberries. *L. elongatus* is especially prevalent in sandy to sandy loam soils, but *L. macrosoma* is more restricted in its distribution, though it is common in soils overlying chalk. *L. attenuatus* Hooper is also of minor importance in southern Britain (Taylor and Brown, 1976). *L. macrosoma* can cause direct feeding damage to the raspberry, but the raspberry is not a host for *L. elongatus*, which does little damage to the crop except as a virus vector.

10.4 NEMATODES AS VIRUS VECTORS

Nematode vectors can acquire and transmit viruses at all stages of their life-cycle. The viruses are ingested with sap from infected plants and selectively and specifically adsorbed onto the feeding stylet or the cuticular lining of the oesophagus: viruses not transmitted by the particular nematode species are not adsorbed (Taylor and Brown, 1981). Thus arabis mosaic virus may be ingested by *L. elongatus* but not absorbed to the specific site of virus retention, but *X. diversicaudatum* feeding on the same infected plant would retain the virus and later transmit it. Viruses do not multiply in nematode vectors nor are they transmitted to offspring (Harrison and Winslow, 1961).

10.5 *MELOIDOGNE HAPLA* AND CONTROL OF NEMATODES

The root-knot nematode, *Meloidogyne hapla* chitwood has been found to multiply on certain raspberry cultivars in North America and it has been suggested that its presence may enhance infection by the crown gall bacterium, *Agrobacterium tumefaciens* (Griffin *et al.*, 1968).

The essential feature of nematode control is to keep the populations to a level that can be tolerated by the crop. The species that attack raspberry have wide host ranges, and so rotations are not usually effective. However, weed-free culture is sufficient to control *L. elongatus*, because the raspberry is such a poor host for it. Before planting a new crop it is desirable to ascertain the level of soil infestation by sampling. Nematodes usually occur in patches and it may be possible to concentrate control measures to these and to prevent the spread of an outbreak of virus disease. Treatment of the soil with fumigant chemicals is the usual method, and although complete eradication is rarely possible the method has been successful in preventing the transmission of virus by *Longidorus* nematodes to raspberry crops subsequently planted (Trudgill and Alphey, 1976). Control of viruses transmitted by *Xiphinema* by soil treatment has proved less reliable, partly because of greater virus persistence in this vector (see p. 119), and partly because the raspberry is a better host for it. Treatment of the soil prior to planting is essential to control the spread of nematode-borne viruses, but for control of *Pratylenchus* it has been demonstrated experimentally that it is possible to take remedial action even after symptoms of infestation become apparent. Applications of systemic nematicides in such circumstances have decreased populations of *P. penetrans* and led to increased yields (McElroy, 1975), though preplanting treatment of the soil is generally best (Trudgill, 1983). A long-term solution is to breed nematode-resistance cultivars. Cultivars differ in susceptibility, and relatively high resistance has been identified in a wild North American red raspberry and in a hybrid of *R. crataegifolius* (Vrain and Daubeny, 1986).

11 The Growth Cycle

11.1 FIRST YEAR'S GROWTH

The biennial growth cycle of most raspberry or blackberry stems starts when a bud from below soil level begins to develop. Two kinds of bud are involved in raspberries: buds from roots, which give rise to root suckers (Fig. 12.4), and basal axillary buds of fruiting canes, which give rise to stem suckers (Fig. 11.1). Root buds arise laterally from uninjured roots, and though they are abundant throughout the year, they only elongate in the "on" season from late summer to spring. Once they start to grow their proximal internodes elongate and carry the growing point towards the soil surface. Some of them travel large distances, but the growing point is well protected by scale leaves. Elongation continues until the soil surface is reached, when the leaves expand to form a tight rosette around the growing point. The leaves are shed and the apex becomes dormant if the shoot emerges in autumn. Hence late-emerging shoots produce little more than a resting bud close to ground level. Buds in leaf axils immediately below the soil become specialized and attain a larger size than those above it. This zone is sometimes referred to as the replacement zone, because the enlarged buds give rise to replacement shoots a year later.

Further elongation of the shoot starts in the spring and continues until the autumn, by which time the shoot may attain a height of 2 to 3 m or more. As it elongates, adventitious roots form freely on its subterreanean part. A second batch of suckers emerges in the spring, but these do not form a rosette like those which emerge in autumn. They grow vigorously and show a continuous transition from closely packed scale leaves below ground to fully expanded leaves separated by long internodes above it. They usually appear within a wide radius of some 20–80 cm from the mother plant, compared with a radius of only 20 cm for the autumn suckers. This gives the raspberry a remarkable capacity to colonize ground over a circle of increasing radius. However, few further suckers are produced from late spring to summer, and the next batch appears in the autumn. Suckers of this batch form a rosette near the soil, cease growth and become dormant (Hudson, 1959; Williams, 1959a).

Canes of primocane-fruiting cultivars can be a metre shorter than those of summer-fruiting cultivars, because the early occurrence of apical flowering

Fig. 11.1. "Replacement" buds just below soil level on a fruiting cane commencing growth to produce new primocanes.

checks their growth, and the earlier the primocane flowers are initiated the shorter the canes will be. Thus when grown at 25°C, "Heritage" canes cease to elongate during the second week of May and flower when they are only 70 cm high; while lower temperatures induce longer internodes, delay flowering and result in taller canes (Ourecky, 1976).

The roots also show a seasonal pattern of growth with a large mid-summer peak. Shoot and root growth begins at about the same time, but root growth continues until much later in the year and is influenced by soil temperature (Atkinson, 1973).

The shoot increases in girth as it elongates. A transverse section of a shoot in early May would show a typical vascular cylinder of pith, xylem, cambium and phloem, surrounded by a ring of longitudinal pericycle fibre bundles, and with a layer of parenchyma and an epidermis outside it. By late May, a periderm develops outside the pericycle and by June four or five layers of periderm cells are present and the secondary vascular tissue is well developed. As growth continues the periderm layer thickens to about 15 cells. The outer cortical layers of cells cannot keep pace with this increase in cane girth, and at various stages longitudinal splits occur.

The first kind of split occurs in late April and continues until June. It is a shallow split involving the epidermis and outer layers of cortical parenchyma to a depth of about two or three cells, and occurs mostly below the nodes but occasionally internodally. A less frequent type of split penetrates radially from the epidermis to the pith, gapes to a V-shaped section and then fills with callus. The most frequent type occurs in the middle and late phases of growth. This type starts near the base of the cane in mid-June and extends rapidly upwards from July to September. It involves the cortical cells, and so the cortex becomes clearly separated from the periderm (Fig. 11.2). The amount of splitting that occurs varies among cultivars but it is a regular

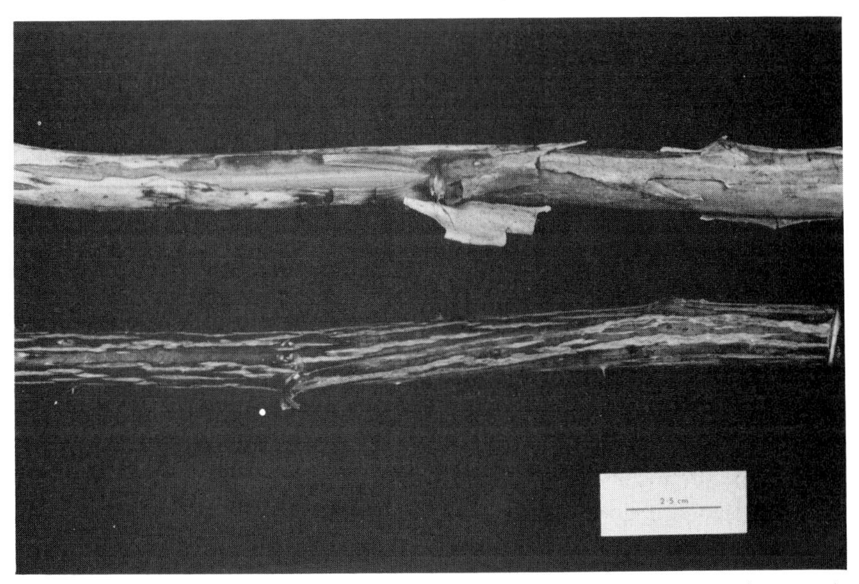

Fig. 11.2. A cane of an *F*1 raspberry × *Rubus crataegifolius* hybrid (bottom), showing characteristic markings caused by the sealing of natural splits in the rind, contrasted with a cane of "Glen Clova" raspberry (top) with a profusely peeling rind. Note the midge blight lesion under a split in the rind on the left-hand part of the "Glen Clova" cane.

feature of cane growth and can be very extensive and result in some 20 to 30 cm of the outer cortex peeling away (Pitcher, 1952). *R. crataegifolius* and its hybrids with raspberry are exceptional in that a wound periderm of suberized and lignified cells forms from the cortex and polyderm as a rapid response to the occurrence of the splits. New tissues from this periderm prevent the peeling of the primary cortex (McNicol *et al.*, 1983). This is important from the point of view of controlling cane midge, which is prevented from completing its life-cycle (see p. 130).

11.2 ONSET OF DORMANCY IN CANES AND SUCKERS

In red raspberries, shortening days and falling temperatures cause the shoot to cease elongation at the end of the growing season. The leaves continue to expand, but in the absence of elongation they form a rosette at the shoot tip. Dormancy sets in but it is a gradual process. It extends over several weeks and is reversible experimentally in the early stages if the plants are returned to long days and high temperatures. Growth is then resumed by the apical bud if dormancy is not too advanced, or by a subterminal bud if exposure to the dormancy-inducing environment was longer.

A stage of complete dormancy is eventually reached which is not readily reversed. Experiments with "Malling Promise" show that this state is reached if the plants are grown for six weeks in a short (9 h) day-length at 10 or 15°C, or even in a 14 h day-length at 10°C, less quickly if they are grown in a 9 h day-length at 15.5°C, but not if they are grown in a 14 h day-length at 15.5 or 21.1°C (Williams, 1959b). It is probably caused by a dormancy-inducing factor formed in the leaves, because prematurely defoliated canes enter dormancy more slowly than canes allowed to shed their leaves naturally (Jennings *et al.*, 1972). Once full dormancy is attained it can be broken by exposing the plants to a period of low temperature, with or without light. A period of six weeks at 4°C is adequate for "Malling Promise", which has a relatively shallow dormancy.

Black or purple raspberries and most blackberries are different from red raspberries both in the time when dormancy begins and in the intensity of dormancy attained. In these fruits, growth continues well into the autumn and the canes do not form a conspicuous terminal rosette of leaves. Growth is less likely to be stopped by the onset of dormancy than by rooting at the tips if the canes are in contact with the ground. Low temperatures also stop growth. However, a period when shallow dormancy is present can be detected. The longer period of growth gives longer canes, but this is desirable for spreading blackberries, because the canes are usually woven along wires between widely spaced plants and most of the canes' length becomes available for cropping in the second year.

In red raspberries, and to a lesser extent in the other *Rubus* fruits, the time when dormancy begins and the intensity that it attains are influenced by the conditions prevailing during the growing season, by the age of the plants and by genetic differences between cultivars. Hence the amount of chilling required to break dormancy cannot be regarded as a constant characteristic of a cultivar. Måge (1975) showed that the dormancy of raspberry canes was more intense if they grew in areas of southern rather than northern Norway. The differences were related to the higher summer temperatures and shorter day-lengths of these areas. It may therefore be significant that irregular bud-break attributed to insufficient winter chilling is most serious in areas such as Israel and Australia, where summer temperatures are particularly high.

Jennings *et al.* (1972) showed that dormancy was less intense in canes of two-year plants than in those of seven-year plants of "Malling Jewel" and "Carnival". Hence canes of the younger plants continued to grow for longer in the autumn and their buds were more capable of starting growth in early winter. For mature plants grown in Scotland, canes of "Malling Jewel" started to become dormant by early September, were fully dormant from early October until the end of November, and then slowly regained a capacity to grow. Those of "Malling Promise" entered rest more slowly and had a shorter period of deep dormancy, while those of "Carnival" became dormant very early and emerged from it early. By contrast, canes of a genotype closely related to the black raspberry did not become fully dormant throughout the winter. In general, canes of early-fruiting cultivars have a less intense dormancy than those of later ones, but canes of all the cultivars studied in Scotland emerge from deep dormancy by mid-December and are then prevented from growing until the spring by a persisting post-dormancy of low intensity, together with the prevailing low temperatures. Some cultivars are prone to enter a state of secondary dormancy during their post-dormancy phase (Jennings, 1987b).

This sequence of dormancy also occurs in the root suckers, which pass the winter as dormant shoots at or just below soil level. The suckers that emerge in September or October acquire a deep dormancy and usually resume growth by an axillary bud close to ground level, while those that emerge later become less dormant and are more likely to resume growth from the original apical meristem, even though they form rosettes. It seems that axillary buds in the replacement zone do not become fully dormant, as they do not need a period of low temperature before they recommence growth, and they frequently grow when the apical bud has had insufficient chilling to break its dormancy. A similar situation occurs when non-dormant plants are planted in spring and receive a cold or drought stress before they are established.

Fig. 11.3. Growth from a subterminal axillary bud taking over from that of the terminal bud, which has become dormant and rosetted.

The check in growth may then be sufficient to induce dormancy in the apical bud and cause a subterminal one to grow (Fig. 11.3).

When raspberries are grown in areas such as Israel, Australia, Mexico and Chile which do not have cold winters, the canes frequently have too few fruiting laterals, because their bud dormancy is not lost by the time temperatures rise in the spring (see p. 162). The choice of cultivar is particularly important in these circumstances. In Chile the problem was not considered too serious in the 1970s, probably because the main cultivar grown then was "Lloyd George", which has a low chilling requirement. In Mexico "Malling Exploit" produced an acceptable number of laterals, and transgressive segregation for chilling requirement occurred in a progeny obtained by open-pollination (Barrientos and Rodriguez, 1980). The problem was more serious in Australia, where the most common cultivar grown was "Willamette", which has a high chilling requirement. D. L. Jennings and G. R. McGregor (unpublished work) surveyed ten cultivars for their fruiting lateral production in Australia, and found that they formed two discontinuous groups, one with late bud-break and few laterals per cane, and the other with early bud-break and good though not always normal lateral production. They found that the first group were still in a state of post-dormancy when the temperature rose in spring. Exposure to wind and other forms of stress aggravated the problem.

An important cultivar in the group with good lateral production was "Marcy", which is one of the most important cultivars grown in New Zealand, and clearly owes much of its success there to its low chilling requirement. Another cultivar in the group was "Glen Clova", which has given poor fruiting-lateral development when grown in Israel (Snir, 1986), possibly because even the chilling needs of cultivars with a low requirement are not satisfied there. Snir (1983) greatly improved lateral production and hastened the cropping season in several cultivars by spraying the canes with 4.0% calcium cyanamide just before the expected time of bud-burst. This and other chemicals appear to reduce the concentration of the dormancy-inducing factor, and hence effectively substitute for chilling.

11.3 INITIATION OF FLOWER BUDS

The initiation of flower buds usually starts at the same time as the canes begin to acquire dormancy, but the two processes can occur independently. For example, flower-bud initiation occurs before the onset of dormancy in primocane-fruiting cultivars, and dormancy occurs without flower-bud initiation in juvenile canes. Factors such as water stress that slow primocane growth in late summer hasten the onset of flower-bud initiation, which also occurs earlier in thin canes than in more vigorous thick ones (Crandall and Chamberlain, 1972). In summer-fruiting red raspberries, flower-bud initiation usually starts in mid-September in Scotland (Robertson, 1957), and slightly later in Oregon (Waldo, 1934). It starts in the terminal bud and in axillary buds five to ten nodes below the apex, and proceeds progressively down the cane until November, when a gap or a period of reduced activity sets in until late-January or February. A second period of flower-bud development then occurs before bud-break starts. Development is slightly later in any secondary buds that are present, and may or may not occur in tertiary buds present in the scale axils of the primary and secondary buds.

Development remains more advanced in the region where it starts, and hence the subterminal buds keep a lead throughout the dormant period. This is seen as an increasing complexity in the floral organs. Within a dormant bud, the terminal apex on the floral axis develops first, and, as this development proceeds, flower primordia arise in succession in the axils of bracts and form the terminal cluster of the inflorescence. Behind this, meristems in the axils of leaf primordia initiate secondary clusters which, in turn, consist of a terminal flower and other flower primordia behind it. Each dormant bud thus contains a complex inflorescence comprising a terminal and several secondary inflorescences. Only the basal apices of the bud remain vegetative.

This sequence occurs later in the buds of other *Rubus* species. Robertson (1957) found that flower-bud initiation started some three weeks later in Loganberries than in red raspberries, but in the black raspberry "Cumberland" it did not start until mid-October. Hence it did not reach an advanced stage in the autumn. Purple raspberries behave like red raspberries in Scotland but like black raspberries in Oregon (Waldo, 1934). In Scotland, Robertson found that the blackberry "Himalaya Giant" started to form flowers in mid-October, but that the blackberry "Ashton Cross" did not start until mid-March. This may explain why the latter has laterals with such a large number of nodes.

Williams (1960) showed that flower-bud initiation in 20-node plants of "Malling Promise" occurred within three weeks if the canes were subjected to 10°C in a 9 h day-length, slightly more slowly if they were subjected to 10°C in a 16 h day-length and not until six weeks if they were subjected to 12.8°C in a 9 h day-length. It did not occur in plants grown at 12.8°C in a 16 h day-length or at 15.5°C whatever the day-length. Most of these conditions also induced dormancy, and it was necessary to subject the plants to a further period of 6 weeks at 4°C to remove it before growth could be induced.

He also found that the age of the "Malling Promise" plants determined the degree to which they responded to an inductive environment of 10°C and a 9 h day-length: plants of only five nodes did not respond at all, plants of 15 nodes variably developed flower buds towards the apex but only after six weeks, plants of 20 nodes always developed flower buds, starting after two weeks, and plants of 30 nodes developed flower buds after only one week's treatment. Six weeks at 4°C without light was sufficient to cause both flower-bud initiation and to break bud dormancy for the last group.

Raspberry cultivars vary in the stage at which their canes initiate flower buds, which seems to occur when the meristem reaches a certain physiological state. The timing of this is determined genetically and is the basis of selection for primocane-fruiting cultivars (see p. 20). Thus Williams (1959a) found the first signs of flower initiation in the terminal buds of "Lloyd George" in late-July, when the shoots were still elongating, and found that it had occurred in all the terminal buds examined by the third week of August, well before it had occurred in the summer-fruiting "Malling Promise".

Environmental factors have large effects, though they probably effect flower-bud initiation indirectly through effects on the plants' overall physiology. For example, Vasilakakis *et al.* (1980) showed for the primocane-fruiting "Heritage" that exposure of the adventitious buds from which the canes grew to low temperature in winter influenced the stage of growth at which flower-bud initiation occurred in the following season, though it was not obligatory for floral initiation. Thus canes from buds that did not receive cold flowered after 80 nodes of growth, but those from cold-treated buds did so after only 41 nodes of growth.

The hormonal factors involved in flower induction are not understood, but it is likely that the varying response of different genotypes to the environment is determined by varying production of indigenous plant hormones. Gibberellins, possibly acting synergistically with cytokinins, may have an important role. Hence Vasilakakis *et al.* (1979a) found that high levels of gibberellic acid were associated with flower induction in "Heritage", but did not discover whether this was the cause or the effect of flower induction. They also found that non-cold-treated plants of this cultivar had much lower levels of cytokinins than cold-treated plants, which suggested that the hormone might be a prerequisite for flowering. Cytokinin activity increases with plant age, which may explain why cold treatment shortens the juvenile growth phase of the meristem, and why mature plants are more responsive to low temperatures and short day-lengths.

High temperature and high nitrogen and other factors that promote cane elongation in spring and early summer also advance the season of primocane-fruiting cultivars and promote the most profuse flowering. For these cultivars primocane fruiting occurs until the axillary buds become dormant, and flower-bud initiation also occurs in the dormant buds below the last inflorescence on the primocane, as in non-primocane-fruiting cultivars (Vasilakakis *et al.*, 1979b).

Tropical species complete all of their growth cycle in short days and high temperatures. The factors which promote their flower-bud initiation have not been studied, but it seems likely that the physiological age of the canes is important.

11.4 ACCLIMATION OF CANES

Acclimation is the third process that occurs in the canes in response to autumn conditions. The cessation of extension growth is probably the key factor which induces it, and the typical sequence leading to acclimation is the cessation of extension growth, the onset of dormancy, and then a reduction in the canes' water content. Genotypes prone to winter injury cease growth and become dormant later than hardy ones; their canes' water content remains high for longer, and their acclimation occurs later. The timing of the acclimation process is influenced by the same factors that influence the cessation of extension growth. It is delayed, for example, in young plants, in plants prematurely defoliated by gales or mite infestation, in plants given excessive nitrogen feeding, and in plants grown in the climatic conditions prevailing in parts of Ireland (Jennings *et al.*, 1964; Jennings and Cormack, 1969; Jennings *et al.*, 1972).

The water content of a hardy cultivar falls to about 55% in October, to about 38% in mid-winter and then rises in March. Although the terminal cane portions retain a slightly higher water content throughout the winter, the spring rise in water content occurs first in the bases of the canes. In red raspberry cultivars which have low winter dormancy, "Malling Promise" for example, the canes may show a very early rise in water content at their cane bases, because they are all able to respond to weather fluctuations during most of the winter. Canes of cultivars with deep dormancy do not show a response until late winter.

These changes in water content are accompanied by movement of food reserves between canes and roots. The starch content of the canes reaches its minimum level in late November and remains relatively low until February, when it rises again, while its sugar content is higher than that of the roots from November to February. The starch content of the roots reaches its maximum in November. Jennings and Carmichael (1975c) considered that these changes were induced by the onset of frosts and not by the onset of dormancy.

The canes normally, but not invariably, shed their leaves during their period of acclimation. Certain blackberries, for example, *R. ulmifolius*, retain their leaves throughout the winter, and raspberry cultivars such as "Glen Prosen" shed their leaves late in the autumn.

11.5 SECOND YEAR'S GROWTH

In the spring of the second year the previous year's vegetative primocanes become fruiting canes. The terminal bud of the primocane is either killed by the winter or removed by pruning, and so, provided that sufficient chilling has occurred, growth is resumed by axillary buds in the upper zone of the cane. These develop into fruiting laterals. Some of the nodes below them remain dormant, but those in the basal replacement zone below soil level elongate vigorously and develop into vegetative replacement shoots (Fig. 11.1). Sometimes basal shoots initiate flowers and produce fruit after the main crop in the autumn, but such basal flowering-shoots should be distinguished from the shoots of a primocane-fruiting cultivar, which initiate flowers whatever their origin.

12 Stems, Leaves and Roots

12.1 STEM NODES AND INTERNODES

Nodes are differentiated on a primocane of a red raspberry at a nearly constant rate throughout the growing season, but the rate of cane elongation varies considerably and is strongly influenced by growing conditions and by cultivar differences. Thus the number of nodes produced in a season is nearly constant, but the number present in the cropping zone (below 150 cm) is relatively low if the canes grow rapidly and become tall, and higher if they grow slowly and reach only a moderate height. Tall canes tend to be thick, and so it is common to have tall, thick canes with few nodes and long internodes alongside shorter, thinner canes with more nodes and shorter internodes. The base and tip regions of the cane have many nodes and short internodes, because they are formed when cane elongation is slow at the beginning and end of the season, while the middle region has fewer nodes because it is formed when elongation is rapid.

When growing conditions are good in Britain, canes of vigorous cultivars produce 200 cm of growth or more during the three summer months, while those of moderately vigorous cultivars like "Glen Prosen" or "Haida" grow just enough to be tipped at 150 cm. After tipping, canes of the vigorous cultivars may have less than 30 nodes and canes of the others have nearly 40 (Jennings and Dale, 1982).

There is a second component of variation that causes thick canes to have more nodes almost regardless of their height. This may be a phyllotaxis effect, because studies of phyllotaxis show that a bud primordium is cut off whenever an apex reaches a certain size and shape. Plant breeders can therefore select for high numbers of nodes within the cropping zone without necessarily selecting for moderate vigour, but they need a statistical procedure to identify and evaluate the two components of variation (Jennings and Dale, 1982).

12.2 AXILLARY BUDS

Most raspberry cultivars have only one bud at each stem node, but some have two or more (Fig. 12.1). In progenies related to "Lloyd George" the

Fig. 12.1. Variation in bud development at nodes of raspberry canes (left to right): primary bud only; primary and secondary buds showing unequal development; and primary and secondary buds showing equal development. The presence of a secondary bud is determined by genes Bd_1 and Bd_2, but their relative development is largely determined by non-genetic factors.

presence of secondary buds is determined by two complementary genes, Bd_1 and Bd_2, whose expression is markedly influenced by their homozygosity, by minor genes and by environmental factors; they are frequently absent from the bases and tips of canes and from weakly growing canes (Keep, 1968c). Many species closely related to blackberries or raspberries show strong secondary-bud development and some have secondary buds present at 90 to 100 per cent of their nodes. By contrast, species of subgenus *Anoplobatus* do not develop secondary buds but have well-developed tertiaries in the axils of their primary bud scales. Clearly, the harsh environments inhabited by *Rubus* species have favoured the development of several systems of lateral replacement to make good the losses incurred from gales, winter cold and spring frosts (Keep, 1969b).

12.3 TIP ROOTING

The canes of black raspberries and blackberries are more indeterminate than those of the red raspberry, largely because they do not become dormant so soon and grow vigorously for longer. Their tips are often terminated by tip roots, which are formed in the black raspberry because of the presence of gene *Tr*, which also confers an ability to form roots on stem cuttings, and hence permits propagation from leaf-bud cuttings (Knight and Keep, 1960).

12.4 STEM MORPHOLOGY

The stems of *Rubus* species are extremely variable, especially blackberry stems. The shape of the stems can be round or angular, they may be spine-free or covered with a varying density of spines and the spines in turn may be of various sizes and shapes. In addition, the stems may be erect, arched or prostrate and they may or may not be pubescent, glandular or glaucus. They are usually biennial, but they are annual in the *Cylactis* subgenus and perennial in some species of the *Anoplobatus*. These and other differences have considerable taxonomic importance, but they are considered here only where they are important in management or disease control.

12.5 SPINELESSNESS AND ITS INHERITANCE

Breeding for spinelessness is a major concern of plant breeders, and there are several major genes which confer it. The first to be discovered was the recessive gene *s*, which Lewis (1939) found in segregates of the Scottish red raspberry "Burnetholm". This gene has been widely used in breeding, and "Glen Moy" and "Glen Prosen" are the first spine-free cultivars to result. Spines are more troublesome in blackberries, and so breeding for spinelessness has higher priority in these fruits. Several genes are being used, but the most important was discovered in the diploid blackberry *R. rusticanus* var. *inermis*, and later shown to be homologous with the recessive gene *s* found in raspberries (see p. 47).

A considerable advantage of gene *s* is that it causes the cotyledons to be free of glands through a pleiotropic effect. This permits selection for spinelessness as soon as the cotyledons emerge from the soil, a considerable advantage for a recessive gene segregating at the tetraploid level. The stems of the spine-free segregates are completely free of spines, but the petioles of juvenile leaves carry a few. A spine-free mutant of "Willamette" raspberry has recently been shown to have the dominent gene Sf_{w}, which has similar effects. The blackberry "Whitford's Thornless" has a different recessive gene for spinelessness.

Dominant genes for spinelessness are more useful for breeding hexaploid or octoploid cultivars, and until recently the only one available was gene *Sf*, derived from the blackberry "Austin Thornless". Seedlings carrying this gene have glandular cotyledons and give rise to plants with a varying density of spines at the bases of their canes, but for practical purposes the canes are spine-free (Jennings, 1984). No cultivars have been bred with the gene, partly because the spine-free segregates have tended to be less vigorous or less hardy than their spiney siblings.

Spine-free mutants such as "Thornless Evergreen", "Thornless Loganberry" and "Thornless Youngberry" have occurred through mutation only in the outer (L1) layer of the stem. The inner layers (L2 and L3) are unaffected, and so the canes are periclinal ("hand-in-glove") chimeras for spinelessness. Since gametes and roots are formed from the inner plant layers, the spine-free mutants produce only spiney progeny and any canes which arise from the roots are spiney. The genes became available to breeders following the discovery of plants in which the L1 layer had ingressed into the inner tissues (Rosati *et al.*, 1986), or in which adventitious buds had arisen during tissue culture of a callus of L1 tissue. McPheeters and Skirvin (1983) used the latter method to obtain material segregating for a dominant gene for spinelessness from "Evergreen Thornless". This gene was later designated S_{TE} (Hall *et al.*, 1986a). Plants carrying it have a few spines at their bases, like those carrying gene *Sf*. Rosati *et al.* (1986) and Hall *et al.* (1986b,c) independently obtained genetic spinelessness from Thornless Logan, which was later found to be conferred by a dominant gene designated Sf_L (Rosati *et al.*, 1988). In this instance the spine-free segregates were entirely spine-free, even on the petioles of their juvenile leaves.

Species of the *Anoplobatus* and *Cylactis* subgenera have spine-free glandular stems and attempts have been made to transfer genes for spinelessness from them to raspberries and blackberries. The first generation hybrids between blackberry or raspberry and *R. parviflorus* of the *Anoplobatus* subgenus are spine-free, but in subsequent backcross generations the development of glands and spines is intermediate between the parental species, and there is evidence that some of the glands develop into soft spines (Jennings and Ingram, 1983). The same is true for hybrids of the raspberry with *R. arcticus* of the *Cylactis* subgenus (Vaarama, 1949). Peitersen (1921) concluded that every glandular hair on a *Rubus* stem is a potential spine, and that spines are often, though not always, developed from glandular hairs. He showed that glandular hairs and spines in *R. allegheniensis* and certain other blackberries were both derived from one to three protodermal cells, and that there were no morphological differences between them except that the terminal cells of the glandular hairs became secretory organs. Nevertheless it is clear that genes of *R. parviflorus* promote glands and suppress spines, and that genes of *R. idaeus* act in the reverse way.

12.6 PUBESCENCE, GLANDULAR HAIRS AND GLAUCUSNESS

Most populations of wild raspberries in Britain are polymorphic for pubescent canes (Fig. 12.2). This is apparently because gene *H*, which

Fig. 12.2. Variation in the morphology of raspberry canes (left to right): glabrous and densely spined (phenotype *hS*); pubescent and sparsely spined (phenotype *HS*); and glabrous and spine-free (phenotype *hs*).

determines cane hairiness, is rarely homozygous because it is linked with a lethal recessive gene (Jennings, 1967a). The hairs are not conspicuous, but they give the canes a dull appearance and prevent them becoming glossy when rubbed. Canes of *R. strigosus* of North America are not hairy, but the widespread use of the European cultivar "Lloyd George" in breeding has produced many hairy-caned North American cultivars. The canes of many blackberries have hairs of varying density, but the inheritance of pubescence has not been studied in these fruits.

Cane hairiness is important because it is associated with resistance to some fungal pathogens and with susceptibility to others. Preference for hairy or non-hairy canes therefore depends on the relative prevalence of different diseases. It is not known whether the effects are due to linkage of gene *H* with genes for resistance or susceptibility to the diseases, or to pleiotropic effects. Gene *H* has pleiotropic effects on spine frequency and size, and it is conceivable that it affects other epidermal characteristics that influence fungal infection.

Glandular hairs have a similar origin to spines and frequently develop into them, but this does not always happen. In *R. phoenicolasius* they persist, and give the stem, leaves and sepals a dense covering of long, glandular structures of variable length (Fig. 12.3). Long and dense glandular hairs are also characteristic of *R. tomentosus* and related blackberries, but the North American red raspberry, *R. strigosus*, has fewer and shorter glands.

A conspicuous waxy bloom on the stems is a feature of the black raspberry and several North American red raspberry cultivars, and it is also associated with a low incidence of certain cane diseases. The appearance of a bloom depends on the physical form of the waxes and is not an indication of the amount of wax present. For example, stems of "Latham" red raspberry have a conspicuous bloom and a large amount of wax, but those of "Malling Exploit" have an inconspicuous bloom even though the amount of wax present is similar. Canes of *R. gracilis* have a particularly conspicuous bloom with 283 μg of wax per cm of surface. This is twice the amount present on

Fig. 12.3. Dense glandular hairs and spines on a stem and petiole of *Rubus phoenicolasius*.

"Latham" and ten times the amount present on canes of "bloom-free" raspberries of genotype *bb* (Baker *et al.*, 1964).

12.7 LEAVES

Variations in the form of *Rubus* leaves are as important taxonomically as variations in the stem. In the red raspberry the leaves of juvenile stems or fruiting laterals are typically three-foliate while those of primocanes are usually five-foliate. They are less divided on male plants of genotype *ff*, whose fruiting laterals have simple obtuse leaves. In other species the leaves vary from the simple three- to five-lobed leaves of *R. crataegifolius*, to the simple but large, palmately-lobed and maple-like leaves with long petioles of *R. parviflorus*. The latter are also densely glandular. In some species there are up to nine-foliate leaves. Most leaf characteristics are expressed in an intermediate way in species hybrids.

In the red raspberry the leaves are glabrous and free of stomata on their upper surface and tomentose with stomata on their lower surface.

12.8 ROOTS

The roots of *Rubus* species are not very variable, but there are nevertheless large differences between the roots of raspberries or blackberries and the

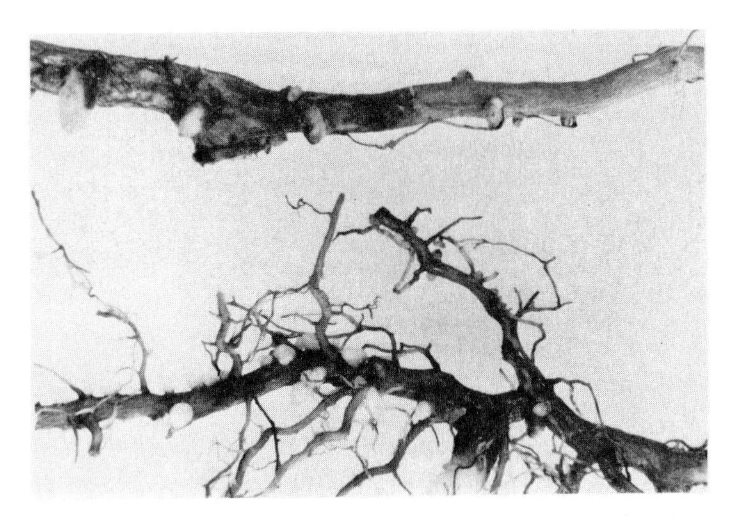

Fig. 12.4. The prolific adventitious buds on raspberry roots in winter which provide a means for rapid propagation.

creeping rootstocks of the arctic raspberries which bear flowering shoots. Roots of red raspberries and some blackberries have adventitious buds which develop in winter on all roots except the thinnest, short-lived feeding ones (Fig. 12.4). The buds appear at irregular intervals on the top, sides and under surface of the roots, sometimes widely spaced, but often spaced densely with about ten buds per cm. A large plant has a very high number of adventitious buds on its roots, but only a proportion of them normally develop into suckers. Some of the remainder will grow if the first flush of suckers is removed to reduce vigour, but the buds of some cultivars produce too few canes in response to this cane-vigour-control treatment. Roots without visible buds can also produce suckers.

Shoots from adventitous root-buds produce suckers if they are retained on intact plants, but they can be used for propagation. Root pieces can be planted directly, or the shoots which arise from them can be cut and rooted as soft-wood cuttings, provided that they are cut in the etiolated zone below soil level. Heydecker and Marston (1968) found that the best length of a cutting was 5 cm for thin roots and 15 cm for thick ones, and that for all thicknesses a planting depth of 1 cm was better than one of 4 cm. They obtained up to ten plants per 30 cm of root.

Knight and Keep (1960) showed that the ability to produce suckers in red raspberry is determined by the recessive gene sk_1 or by the complementary genes sk_2 and sk_3, which are epistatic to $Sk_1\ sk_1$ when homozygous. Black raspberries and blackberries do not have these genes and do not produce suckers.

13 Lateral Branches and Fruiting Laterals

13.1 VEGETATIVE BRANCHES

Apart from cultivars specially selected for fruiting on primocanes (see p. 19), canes of biennial *Rubus* species produce their fruit in the second year on fruiting laterals which develop from the axillary buds of the first-year's primocanes. Not all lateral branches develop on fruiting canes, however, and vegetative branches on primocanes are a feature of newly established plants of vigorous cultivars. These augment the fruit-bearing surface of young plants at a time when they are not always capable of producing adequate numbers of primocanes. Such branching is minimal or absent in mature plants, but it is strongly expressed by certain species closely related to the raspberry, for example, *R. coreanus*. In this species, two dominant genes, designated Br_1 and Br_2, confer a branching habit which segregates in families derived by crossing it with the raspberry (Keep *et al.*, 1977c). In black raspberries and some cultivars of blackberries the primocanes are commonly pruned in mid-season to promote the growth of vegetative laterals and produce a compact plant upon which a second series of laterals bears fruit in the following year. Laterals produced in the second year are vegetative or "blind" if the primocanes are not exposed in the autumn to an environment capable of inducing flower-bud initiation: the amount of exposure required is a varietal characteristic.

Primocane-fruiting genotypes whose canes naturally form flowers in their first year may be particularly prone to form vegetative branches, especially if they are early flowering. This is a valued feature which increases the plants' fruit-bearing surface (see p. 21), but this branching is associated with the physiological state of a flowering primocane and should not be compared with truly vegetative branching. The occurrence of branches is highly correlated ($r = 0.76$) with the occurrence of early autumn flowers (Lawrence, 1976).

13.2 NUMBERS AND POSITION OF FRUITING LATERALS

The number of fruiting laterals present on canes of summer-fruiting cultivars depends partly on the number of nodes in the cropping zone (see p. 152), and partly on the proportion of them whose buds develop into laterals. Each of these components is influenced by cultivar and environmental factors. On average, only about two-thirds of the nodes develop fruiting laterals, largely because the apical dominance of the upper buds limits the development of the lower ones. Hence the probability of a node bearing a fruiting lateral is related to its position on the cane, but other factors interact with apical effects in some instances.

This is illustrated by canes which have not received adequate chilling (see p. 148): these typically have a few terminal laterals, a lateral-free subterminal zone and bear most of their laterals towards the base. The lateral-free zone may extend for a few nodes or for the greater length of the cane. Similarly, in a survey done in Scotland and British Columbia, the proportion of lateral-bearing nodes recorded was slightly lower than predicted just below the cane tip and higher than expected at the eighth node below it. It seems that the subterminal position is prone to stress factors which interact with the apical dominance gradient of the cane (Jennings, 1987b).

Thick canes tend to have fewer fruiting laterals, but they are preferred because they are associated with high yield per lateral and high total yield (Crandall *et al.*, 1974b).

13.3 MULTIPLE FRUITING LATERALS

Only one fruiting lateral develops at each node in most cultivars. It is usually derived from a primary bud (Fig 12.1); secondary buds, if present, usually remain dormant unless the primary bud or lateral is injured. If this happens, secondary buds can produce strong replacement laterals capable of bearing an adequate crop. If strong secondary buds are lacking, one or more lateral replacements may arise from tertiary buds, but these are usually weak and spindly and do not crop so well.

In some cultivars secondary laterals develop and make a useful contribution to fruit yield even when the development of the primary lateral is not interrupted (Fig. 13.1). In "Lloyd George", for example, the average number of laterals per cropping node may be as high as 1.28 (Wood and Robertson, 1957) and values of up to 1.50 have been recorded for "Glen Clova". Even higher numbers have been recorded for a wild-raspberry selection (Knight, 1986). A secondary bud must be well developed to compete successfully with an uninjured primary, and the frequent development of two cropping laterals at a node in a cultivar like "Glen Clova" depends on many factors, including plant vigour and cane diameter, for which the optimum is about 1 cm. Provided that the cane is adequately furnished with secondary buds, the ability to produce multiple laterals from

Fig. 13.1. Multiple fruiting laterals at a node on a cane of "Glen Clova" raspberry.

them is inherited in an additive way as a continuously varying character, and variations in cane diameter are inherited as a component of this variation (Jennings, 1979c).

13.4 LENGTH OF FRUITING LATERALS AND THE NUMBERS OF FLOWERS AND FRUIT PRESENT

The process of flower-bud initiation begins in buds just below the apex of the cane and proceeds both upwards and downwards. The onset of flower initiation in the terminal meristems of these buds prevents further differentiation of nodes within the bud, and so the total number of lateral nodes differentiated is influenced by variations in the time when initiation starts: an early start limits it and a late one prolongs it (Williams, 1959c; Jennings, 1964c). This in turn affects the length of the lateral but not necessarily the number of flowers that it eventually produces. It partly explains why Dale (1979) found that lateral length and the number of nodes present on a lateral was usually greater for laterals towards the cane base, though in some cultivars he found that the laterals were relatively short at the base and showed a curvilinear relationship with node position. He showed that reproductive characteristics such as the number of flowers and fruit present on the lateral tended to be constant for all node positions, though there were exceptions, and all lateral characters were influenced by management and plant age. For example, reduction of competition from young canes in spring increased the length and potential yield of the middle and lower laterals of the fruiting canes. Indeed, fruiting laterals on the lower half of the cane but not those on the upper half have a remarkable capacity to compensate for losses of laterals by producing more fruits of higher individual weight. Losses of up to 50% of the laterals can be tolerated with no reduction in yield per cane (Braun and Garth, 1984).

The number of fruit borne on laterals varies considerably, partly because of variation in the number of flowers initiated and partly because ripe fruit are usually obtained from only a proportion of the flower buds present: fruit development may be arrested at any stage, sometimes as green flower buds, sometimes as flowers and sometimes as immature fruits. Dale and Topham (1980) showed that much of the variation in lateral characteristics could be conveniently considered under three headings: general vigour, which includes lateral length, reproductive vigour and unachieved reproductive potential. The latter includes the buds and flowers which fail to develop into ripe fruit. Good management based upon primocane-vigour control increases the numbers of fruit produced by increasing reproductive vigour (promoting the development of more flowers) and by reducing unachieved reproductive potential (reducing the proportion of flowers whose development is arrested). In general, management and environmental factors influence these developments because they affect the amount of carbohydrate available per fruiting lateral: for example, for thick and thin canes of "Puyallup", the mean supply of carbohydrate per lateral was found to be

143.5 and 83.5 mg of glucose equivalent respectively, and the corresponding average numbers of fruit present were 18.8 and 14.9 (Crandall *et al.*, 1974a).

Genetic variation in fruit numbers per lateral probably reflects differences in the utilization of assimilates by the plant as a whole, as well as differences in the morphology of the fruiting laterals. Among cultivars there is variation in the number of nodes present on the lateral and in the proportion of them which bear flower buds, while within breeding material there are genotypes which produce very large numbers of flowers, up to 10 per lateral node in one reported example, and often 30 to 50 per cent more than are capable of developing into fruit. This represents considerable unfulfilled potential for fruit production, and presents a challenge for breeders to find genotypes capable of harnessing it more effectively (Dale, 1977; Jennings *et al.*, 1976).

Another way to increase the fruit numbers of raspberry laterals is to transfer genes from closely related species such as *Rubus cockburnianus*, which have as many as 100 very small fruit per fruiting lateral, and then to select for intermediate numbers of better size fruit during repeated backcrossing to the raspberry. From crossing with *R. cockburnianus*, Keep and Knight (1968) obtained first-generation hybrids with averages per lateral of from 31 to 73 fruits, and advanced backcross hybrids with numbers of up to 35 were eventually obtained (Knight, 1986). In some instances fruit size had been restored almost to that of commercial cultivars by the second backcross.

In contrast to the fruiting laterals of red raspberries, those of black raspberries have a characteristic terminal cluster of fruit, due in part to the short lateral internodes in this region and in part to the frequency of lateral nodes bearing several fruit. The fruiting laterals of most European blackberries are also different in that they are longer and have more nodes, each with a well differentiated cluster of fruits. Consequently, 50 fruit are frequently found on laterals of European blackberries, but "Darrow" and certain other cultivars from North America have short, early-flowering laterals rarely bearing more than 10 fruits. These laterals flower early, because they start to grow early and because little lateral growth occurs before flowering starts (Fig. 4.1).

14 Flowers and Fruit

14.1 SEPALS AND PETALS

A raspberry flower has five sepals and petals. The sepals, which persist until the fruit is ripe, may be modified in size and number by certain major genes, for example the recessive gene *d*, first described by Lewis (1939) and later redesignated sx_3 by Keep (1964). This gene induces a condition known as sepaloidy, whose effect ranges from the addition of an extra whorl of sepals with the normal number of petals but fewer anthers, to two extra whorls of sepals with either both petals and anthers completely suppressed, or with just a few anthers present and modified towards a sepaloid or carpeloid condition; overall the number of sepals may range from 5 to 17. Another major gene with a major effect is L_1, which gives very large sepals whose lobes are contracted into a narrow point. In the raspberry, the sepals reflex away from the fruit, but in some other species of the *Idaeobatus* subgenus, for example *R. phoenicolasius* and *R. crataegifolius*, they close after pollination and completely enclose the fruit until it is ripe, when they open again.

The petals of a raspberry are small and white but in many *Rubus* species they are various shades of pink. In *R. coreanus*, another species of the *Idaeobatus*, this is due to the presence of gene *An* (Keep *et al.*, 1977a), while in *R. parvifolius*, also of this subgenus, it is due to a series of allelic genes which confer varying intensities of pink colour (D. L. Jennings, unpublished work). In blackberries the petals are generally larger, and in *R. macropetalous*, a dioecious western American species, the petals of the male flowers are 50% larger than those of the female ones. The petals are much enlarged in the *Anoplobatus* subgenus, and provide the attractive display which earns species of this subgenus their popularity as ornamentals; examples are *R. parviflorus*, the thimbleberry of western America, whose petals are white and up to 6 cm across and *R. odoratus* whose petals are of similar size but an attractive deep purple.

14.2 STAMENS AND CARPELS

Raspberry stamens arise in two crowded whorls in numbers ranging from 60 to 90, a smaller range than that shown by the styles.

Table 14.1. Mean numbers of stamens and styles in flowers of three forms of "Malling Jewel" raspberry.

	Diploid	Tetraploid	Diploid gene *L*	S.E. means
Stamen numbers	87.20	82.00	74.11	2.62
Style numbers	74.40	68.40	112.20	3.81

The numbers of both stamens and styles in the raspberry are affected by ploidy, by genotype and by major genes. In "Malling Jewel", for example, a tetraploid mutant and a diploid somatic mutant carrying gene L_1 both have larger and fewer stamens than the diploid from which they arose; the styles also tend to be larger in the two derived forms, but whereas the number of styles is reduced in the tetraploid it is considerably increased in the mutant, where the receptacle is elongated into an enlarged cone. Average counts obtained for each form were as given in Table 14.1 (Jennings, 1966a). Some other important genes in the raspberry are gene *M* and gene *F*, whose segregation can confer a dioecious habit, the recessive gene *f* suppressing the development of female parts of the flower and the recessive gene *m* suppressing the male parts. Most raspberries have genotype *FM* and are hermaphrodite, but *fM* genotypes are male, *Fm* genotypes are female and *fm* genotypes are neuter.

The styles arise spirally on the terminal part of the receptacle, whose size and shape consequently determines the size and shape of the fruit. It is flat in species of the *Cylactis* subgenus, a rounded dome in *R. strigosus* and *R. occidentalis* and a more elongated dome in *R. idaeus* and many blackberries. Selection for larger fruit has given larger receptacles furnished with more styles: whereas in older cultivars the number of styles and hence the potential number of drupelets is rarely above 60, in "Krupna Dvoroda", a large-fruited cultivar, it is often 160. (Mišić *et al.*, 1972.)

14.3 POLLEN AND NECTAR

Pollen grains vary in size and pore number: two variables frequently used to determine their probable ploidy. The relationship between ploidy and pollen size was well illustrated for European blackberries by Gustafsson (1943). Similar relationships exist for raspberries, and the largest grains of the genus are found in the octoploid and duodecaploid blackberries of the *Ursini* section. Pore numbers give a similar series. Diploid raspberries and blackberries usually have three pores, though the range is from one to six, while tetraploids usually have four with a range of from two to six (Zych *et al.*, 1968).

Raspberries and blackberries both secrete nectar profusely. Individual flowers of blackberries, for example, may secrete 14.5 mg of sugar in a diurnal rhythm during their 90 hours of existence (Percival, 1946) and raspberry flowers may yield up to 33 mg of nectar containing about 50% of sugar (Petkov, 1963).

14.4 FRUIT STRUCTURE AND DEVELOPMENT

After fertilization each ovary develops into a drupelet, which can be regarded as a miniature drupe. A drupe fruit is defined as one which develops entirely from a single ovary, and *Rubus* fruits are aggregates of drupelets formed by the ripening together of a number of ovaries all from the same flower and adhering to a common receptacle: in a sense each drupelet is a complete fruit in itself and a miniature homolog of such drupe fruits as the cherry, plum or peach.

The cohesion of these drupelets depends on the entanglement of epidermal hairs, which are unicellular linear trichomes arising from surface cells. The hairs are abundant at the bases and sides of the drupelet, and shorter and less profuse over its dome; they are so tightly enmeshed that the drupelets cannot normally be separated without tearing the skin. In the red raspberry cultivars so far examined they appear to be the only cause of drupelet cohesion, but in some black raspberry cultivars the centres of the commissural sides of the drupelets are also held by a weak fusion of the cuticle or wax on the outer surface of their epidermal cells. In blackberries the hairs are less profuse and the drupelets consequently less compressed than in raspberries. However, variation in drupelet cohesion in raspberries depends more on variation in the percentage set of the drupelets (see p. 121).

Internally, the soft part of the drupelet consists of thin-walled parenchymatous cells, radiating from the pyrene in the centre to a region of larger, oval-shaped cells underlying the epidermis and a slightly collenchymatous hypodermis of one to three cells. The oval cells occupy a comparatively larger part of black raspberry fruits. Fruits are succulent because the walls of the parenchyma are thin and the cells turgid. Cultivar variations in fruit firmness are probaby determined by variation in the frequency of cells of small diameter, in tissue compactness and in overall cell size, which influences the amount of supporting cell-wall surface (Reeve, 1954; Reeve *et al.*, 1965). Barritt *et al.* (1980) measured fruit firmness with a pressure gauge and found that fruit of "Glen Prosen" was some four times as firm as that of soft-fruited cultivars.

Raspberry fruits usually ripen in from 30 to 36 days after pollination and all their drupelets usually ripen together. The small variations which occur are usually determined by the genotype of the maternal parent or by environmental factors, but up to six days increase in the time required for ripening may occur through scarcity of pollen (Jennings, 1971a; Jennings and Topham, 1971). In blackberries the ripening time varies from 40 to 70 days, depending on genotype and temperature (see p. 56). Fruits of both raspberries and blackberries follow the pattern of other drupeaceous fruits in showing three distinct phases of development. Following pollination there is a period of rapid growth due to cell division; this is followed by a period of slow growth during which the embryo develops and the endocarp becomes hardened, and finally there is a period of very rapid growth due to cell enlargement. Each stage lasts 10 to 12 days in the raspberry (Hill, 1958).

14.5 FRUIT RIPENING

The physiology of fruit ripening is slightly different in raspberries and blackberries (Blaupied, 1972; Walsh *et al.*, 1983; Burdon, 1987). Ethylene production in the raspberry rises as soon as the fruit begins to change colour and reaches a maximum when it is fully ripe. Thus the raspberry is a typical "climacteric" fruit, except that respiration, measured as carbon dioxide produced per gram of fruit, decreases as ripening proceeds. By contrast, ethylene production in the blackberry begins at a late stage of colour development in Maryland, U.S.A. and not at all in Scotland. This conclusion arises from studies of four blackberry cultivars in Maryland, where the fruit ripens in mid-summer, and of five different cultivars in Scotland, where the fruit ripens much later. Hence the difference may be due to differences in cultivar or in temperature at ripening time. However, it is interesting that in the Scottish study the behaviour of the blackberry–raspberry hybrids "Bedford Giant", Tayberry and Tummelberry was intermediate between those of the raspberry and blackberry. In the ripening raspberry, the ethylene is produced in equal amounts by the receptacle and the drupelets, and there is a peak of production at the time of petal abscission. Cultivars differ in the amount of ethylene that they produce and there is a tendency for cultivars with high production to have easy fruit abscission. For blackberries grown in North America the ethylene-releasing chemical "Ethrel" has been used as an aid to ripening.

14.6 FRUIT ABSCISSION

Abscission layers form in the tissues as the fruit ripens. In blackberries a single layer forms at the base of the receptacle, which separates from the plant with the fruit, and in raspberries a large number form, one at the point of attachment of each drupelet to the receptacle, which is retained on the plant after fruit abscission. Blackberry–raspberry hybrids are heterozygous for both types of abscission and often do not absciss easily by either method, though most of them absciss like their blackberry parent. Two modes of cell separation occur within raspberry abscission zones. The existing intercellular spaces in the abscission zone first develop into large cavities, owing to growth stresses and a combination of fracture of the middle lamella and partial dissolution of the walls of the cortical cells. These cortical cells remain intact, but in the vascular bundle the development of cavities is followed by cell wall breakdown and cell disintegration (Mackenzie, 1979). These processes are gradual: hence the force required to remove underripe fruit of "Glen Clova" or "Malling Jewel" falls from about 300 g for underripe fruit to 25 g when the fruit are overripe.

14.7 FRUIT COLOUR

14.7.1 The anthocyanin sugars

The anthocyanin molecule consists of an aglycone with a varying number of sugar residues attached to the hydroxyl group in the 3 position. The aglycones of *Rubus* are cyanidin and pelargonidin and the number of sugar residues attached varies from one to three. The monosaccharide form is cyanidin-3-glucoside and additional glucose, rhamnose or xylose sugars may be present in various combinations to give diglycosides or triglycosides. Thus an additional glucose gives the diglycoside known as cyanidin-3-sophoroside, an additional rhamnose gives the diglycoside known as cyanidin-3-rutinoside and an additional xylose gives the diglycoside known as cyanidin-3-sambubioside. The addition of both glucose and rhamnose sugars gives the triglycoside cyanidin-3-glucosylrutinoside (Fig. 14.1). A more rare diglycoside is cyanidin-3,5-diglycoside, which occasionally occurs in red raspberry fruits; in this one the second glucose combines with the aglycone in the 5 position. No special enzyme systems are required for the synthesis of triglycosides: they are formed whenever the two appropriate diglycoside synthesizing systems are present together. These are controlled by the genes *R*, *So*, and *Xy* which confer the ability to synthesize rhamnose, sophorose and xylose sugars respectively (Jennings and Carmichael, 1980).

14.7.2 The aglycones

Variation in the aglycone part of the anthocyanin molecule also occurs in *Rubus*: cyanidin pigments are the most common but pelargonidins are also

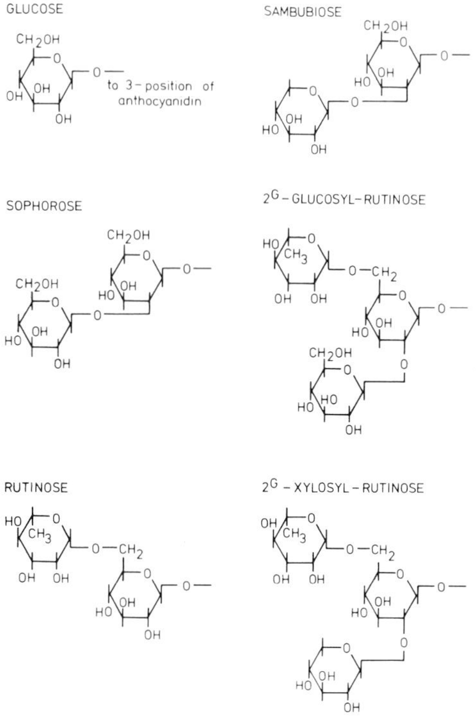

Fig. 14.1. Chemical structure of the six common anthocyanin pigments of raspberry: cyanidin glycoside contains a simple glucose sugar; the diglycosides cyanidin-3-sophoroside, cyamidin-3-rutinoside and cyanidin-3-sambubioside contain in addition a second glucose, rutinose or xylose; while the triglycosides cyanidin-3-glucosylrutinoside and cyanidin-3-xylosylrutinoside, respectively, contain a rutinose together with a sophorose and a rutinose together with a sambubiose (Jennings and Carmichael, 1980).

found, usually as traces alongside the cyanidins but sometimes as major pigments, for example, in *R. pileatus* and *R. crataegifolius* of subgenus *Idaeobatus* and in *R. parviflorus* of subgenus *Anoplobatus*. Variation in the sugar residues attached gives six pelargonidin pigments corresponding to the six cyanidins. Fruit with a predominance of pelargonidin glycosides tend to have an orange-red rather than a true red colour.

The kind of pigments present is a diagnostic characteristic of many of the fruit types. For example, both red and black raspberries have the monoglucoside form, cyanidin-3-glucoside, together with the diglycoside cyanidin-3-rutinoside, but only red raspberries have the diglycoside cyanidin-3-sophoroside and the triglycoside cyanidin-3-glucosyl-rutinoside, and only black raspberries have the diglycoside cyanidin-3-sambubioside and the triglycoside cyanidin-3-xylosylrutinoside. All six forms are found in first generation hybrids, because the hybrids have a full complement of the genes *R*, *So* and *Xy*; exceptions occur because some red raspberries do not carry gene *R* which controls the formation of rhamnose-containing glycosides.

The blackberries so far examined, whether from the *Eubatus*, *Orobatus* or *Malachobatus* subgenera, contain a very limited range of pigments. Usually cyanidin-3-glucoside predominates and cyanidin-3-rutinoside is sometimes present in smaller quantities. Anthocyanins characteristic of red raspberries are found in "Bedford Giant" and in other hybrid berries which have red raspberries in their ancestry, while the fact that the Andean blackberry, *R. glaucus*, whose classification is difficult, has xylose pigments, strongly supports the view that it is related to the black raspberry.

14.7.3 Anthocyanin concentration

The concentration of anthocyanins in the fruit is determined by another series of genes; gene *T* is the most important of them, its recessive allele *t* giving a very low concentration of anthocyanins. This allele affects all parts of the plant, giving fruit with yellow colour and stems with non-pigmented spines.

A recessive inhibitory gene, *i*, found in a Russian wild raspberry, interacts with gene *T* when it is homozygous to give apricot-coloured fruits and leaves and stems pigmented to varying degrees (Keep, 1984). In the presence of the intensifying gene *P*, *tt* genotypes do not block anthocyanin synthesis so completely and the higher pigment concentration attained gives red-tinged spines. Crane and Lawrence (1931) considered that the gene also gave apricot fruit colour, but this was not supported by the results of Keep (1984). Another gene (B1) has been postulated to determine black fruit colour in the black raspberry, the purple colour of the red-black hybrid being due to its heterozygous condition (Britton *et al.*, 1959). However, this interpretation is not supported because fruits with a continuous range of colour intensities segregate in advanced generations of red × black raspberry hybrids.

A dominant gene designated *Y*, found in *R. phoenicolasius*, also causes reduction in fruit pigment concentration and gives yellow fruit colour but, unlike gene *t*, it has no effect on the pigments of vegetative growth. Its effect is suppressed in *R. phoenicolasius* itself by gene *Ys*, with the result that the species has red fruits. Gene *Ycor* has the same or similar effect to gene *Y* and occurs in *R. coreanus*. It is interesting that all these genes limit the concentration of the fruit pigments without preventing the development of any one anthocyanin in particular (Jennings and Carmichael, 1980).

14.7.4 Colour expression

Several chemical factors influence the colour of anthocyanin solutions, but pH is probably the most important of them. This is possibly because anthocyanins are chemical indicators that are red in acid solutions, violet or purple in neutral solutions and blue in alkaline ones. Jennings and Carmichael (1979) suggested that this property could explain why the colour of blackberries frequently changes from black to red when the fruit is frozen: freezing causes widespread cellular disruption which possibly allows mixing of the cells' plasma and vacuolar contents and places the anthocyanins in a solution of lower pH than where they occur in non-frozen fruit.

Thoroughly ripe blackberries do not become red when frozen, and the problem arises because black fruit are not always fully ripe. This is probably because anthocyanins start to develop first within cells near the skin of the fruit, where they are exposed to sunlight (Grisebach, 1982). Consequently, slightly immature fruits cannot be recognized until after freezing, when visual discrimination is possible. These immature fruit also have a lower concentration of anthocyanins, and a more likely explanation of the colour change is that the cell disruption allows the anthocyanins to spread from the vacuoles, where they are mostly concentrated, and to become diluted by the cytoplasmic sap. For the immature fruit this pigment dilution is in itself sufficient to cause the colour change (Polesello *et al.*, 1986).

14.8 FRUIT CHEMISTRY

The main constituent of both raspberries and blackberries is water. In addition raspberries contain about 14 per cent of solids, of which about 9 per cent are soluble and 5 per cent are insoluble; blackberries tend to have a higher percentage of insoluble solids owing to the greater size and weight of their pyrenes. Pectins are an important constituent of the soluble solids, but raspberries have a low content of these substances, containing an average of only 0.4% (w/w expressed as calcium pectate) with a range of from 0.10 to 0.97%. Blackberries average 0.8% with a range of from 0.35 to 1.19% and Loganberries have an intermediate content (Green, 1971). The protopectins form the intercellular cement and contribute towards the firmness of fruit texture. The amount of pectic substances present decreases with ripening, largely due to hydrolyses of the protopectins, but Duclus and Latrasse (1971) found that the hydrolysis was relatively greater in the soft-textured "Malling Exploit" than in the firmer-textured "Puyallup".

Table 14.2. The sugar content of mature *Rubus* fruit (Green, 1971).

Fruit	Sugars (% w/w)			
	Reducing	Sucrose	Total	Ratio sugar/acid
Blackberry	—	—	4.3	2.8
Boysenberry	4.20	1.14	5.34	3.5
Loganberry	—	—	3.4	1.3
Raspberry min.	1.44	0.06	1.57	0.9
Raspberry max.	4.25	1.21	5.34	0.9

14.8.1 Sugars, acids and volatiles

The flavour of raspberries and blackberries is determined by their content of sugars, acids and volatiles, which varies considerably with variety and growing conditions. It is not strongly correlated with any one of them, because it is not perceived as a number of separate characteristics but more as an overall impression. Generally, fruit grown in areas which have warm dry summers have a higher content of sugars and are more aromatic and more highly coloured than fruit grown in more humid and milder regions.

The main sugars are the reducing sugars glucose and fructose, which are present in approximately equal quantities, and sucrose. They contribute the major soluble component of the juice. The organic acids contribute the second biggest contribution to the soluble component, and the ratio of sugar to acid plays a part in determining the level of flavour acceptance (Table 14.2). The main acid of raspberries is citric acid with very little malic acid present; in blackberries citric acid is of little or no importance and malic acid and iso-citric acid with its lactone predominate (Green, 1971). This is shown in Table 14.3.

Table 14.3. The major acid content of *Rubus* fruit (Green, 1971).

Fruit	pH	Acidity calculated as % w/w				
				Individual acids		
		Typical ripe fruit	Range	Citric	Malic	Iso-citric and lactone
Blackberry	3.00	1.50	0.68–1.84	Tr.	0.82	0.81
Boysenberry	2.91	1.51		1.24	0.21	0.06
Loganberry	2.90	2.63	1.02–3.12	2.10	0.53	—
Red Raspberry	3.47	2.80	0.74–3.62	2.06	0.80	—
Black Raspberry	3.56	0.90	0.84–1.01	—	—	—

The acids have a high buffering capacity which imparts stability to the pH, keeping it close to 3.0. Titratable acidity therefore gives a more precise measurement of the quantity of acid present. As fruit development proceeds, this quantity increases at first and then decreases as the fruit begins to ripen (Hill, 1958). It is lower at high temperatures. At this stage the relationship between titratable acidity and ripeness is so close that it can be used as a quantitative measure of fruit ripeness. This has proved useful for studies of the efficiency of machine-harvesting regimes in both blackberries and raspberries (Kattan *et al.*, 1965; Mason, 1974). There is also evidence that the proportion of the different acids changes with ripening. For example, in the Boysenberry, the proportion of malic acid was found to decrease from 17.9 to 8.3 per cent of the total acid content as ripening proceeded, while the proportions of citric and iso-citric acids increased. There are also at least ten trace acids in raspberries and a smaller number in blackberries, including shikimic, quinic, mucic and lactoisocitric (Green, 1971).

The advent of gas chromatography has led to the isolation and identification of a very large number of volatile compounds that probably contribute to the odour and flavour of fruits; they are present in complex mixtures and their total content is no more than 10 p.p.m. Those identified in raspberry fruit include large numbers of alcohols, hydrocarbons, carbonyls, esters, ketones and some miscellaneous compounds (Nursten and Williams, 1967). The nearest to an odour impact compound that has been identified is the "raspberry ketone" or 1-(p-hydroxyphenyl)-3-butanone, and important contributions are made by *cis*-3-hexen-1-ol, which has a fresh, grassy odour, α- and β-ionone and α-irone (Nursten, 1970). The alpha and beta ionones were among the volatiles thought to contribute most to flavour and scent by Hiirsalmi and Säkö (1976). However, Kallio (1976a,b) identified 200 compounds from the fruit of *R. arcticus* and found that the compound which gave this fruit its character was mesifurane, the content of which rose from 0.2 to over 20% of the volatiles during ripening. The aroma of all raspberries is very sensitive to heat and deep freezing, which destroy the aroma nuances present in fresh fruit.

In France, a procedure has been developed to obtain an "aroma index" for assessing raspberry flavour. Selections are first screened for the required sugar and acid contents, and the steam distillates of 50 g samples from those selected are treated with vanillin in concentrated sulphuric acid. Aroma indices are then determined following a standardized colorimetric procedure. The results are highly correlated with those obtained from sensory assessments (Latrasse *et al.*, 1982), and the following six major compounds were found to account for 80% of the aroma-index value of the aromatic cultivar "Malling Admiral": geraniol, linalol, nerol, α-terpineol, and α- and

Table 14.4. The vitamin content of *Rubus* fruits (mg/100 g) (Green, 1971).

	Vitamin A	Vitamin B			Vitamin C
Fruit	Carotene	Thiamine	Riboflavin	Nicotinic acid	Ascorbic acid
Blackberry	0.10–0.59	0.03	0.034–0.038	0.4	20
Boysenberry		0.02	0.13	1.0	13
Red Raspberry	0.05–0.08	0.02–0.03	0.03 –0.09	0.4–0.9	25
Black Raspberry	Tr.	—	—	—	18

β-ionones. Comparison of aromatic and non-aromatic cultivars supported the hypothesis that the aroma was due to monoterpene alcohols, especially geraniol and linalol, and to the α- and β-ionones (Latrasse, 1982).

Raspberry fruit also contains 0.10 to 0.14 per cent of tannins, including the hydroxy acids chlorogenic, ferulic and neochlorogenic and also catechin and the anthocyanins (Green, 1971).

14.8.2 Vitamins, minerals and proteins

Raspberries and blackberries are poor sources of vitamins (Table 14.4). Only vitamin C is present in appreciable amounts, but even the content of this vitamin is only a tenth of that of the black currant (Green, 1971). Similarly, the mineral content is low (Table 14.5), and though it varies with

Table 14.5. The nitrogen, protein and mineral content of *Rubus* fruits (mg/100 g) (Green, 1971).

	Blackberry	Red Raspberry	Black Raspberry
Total N	181	177	—
Protein	0.56	0.50	0.39
Amino acids (mg/100 g)	2.25	1.99	1.41
Ash	0.50	0.50	0.60
P	23.8	22.0	22.0
K	208.0	168.0	199.0
Na	3.7	1.0	1.0
Ca	63.3	22.0	30.0
Mg	29.5	21.0	—
Fe	0.85	0.9	0.9
Cu	0.18	0.21	—
S	9.2	17.3	—
Cl	22.1	22.3	—

cultural factors the influence of the latter is not as great as might be expected. Potassium and calcium are the most important. The nitrogen content is also low, but the fruits do contain proteins, polypeptides and traces or larger amounts of some 15 amino acids. Chromatographic maps of the latter have been suggested as a possible means for determining the authenticity of *Rubus* fruit juice.

15 Pyrenes and Seeds

15.1 PYRENE STRUCTURE AND SIZE

In the centre of each drupelet is a pyrene with a hard endocarp enclosing one or occasionally two seeds. The endocarp consists of two layers of elongated, parallel scleroid cells at right angles to each other and each several cells thick, the outer one forming the characteristic ridged pattern of the pyrene (Reeve, 1954). This pattern has been used to identify *Rubus* seeds in fossils, and four types of pattern have been recognized among Japanese species (Satomi and Naruhashi, 1971). Curiously, although the endocarp is entirely maternal in origin, its size and shape is considerably influenced in the raspberry by interactions between the pollen parent and the maternal tissues (Jennings, 1971b).

The two qualities of the pyrenes which contribute most to fruit and jam quality are their size and their tendency to become "blind" in jam. During storage of raspberry jam, a slow displacement of air from within the pyrenes allows the surrounding syrup to infiltrate, causing them to lose their opacity and become "blind". They appear to have merged with the surrounding jam, which acquires a dull, old appearance and gives the illusion of a low fruit content. The rate at which this happens is primarily determined by the identity of the raspberry cultivar, but it is aggravated by certain manufacturing procedures, and proceeds particularly quickly in jam made from sulphited fruit.

Small pyrenes are preferred for fresh fruit and processed products. This is especially so for blackberries whose pyrenes are usually much larger than those of raspberries. Surveys have shown a wide range, however. Among some American blackberries, mean pyrene weight, which is indicative of size, ranged from 2.01 to 4.83 mg, while in breeding material in Scotland it ranged from 2.10 to 6.10 mg. In both raspberries and blackberries the character is inherited quantitatively; dominance tending towards large pyrenes in raspberries and towards small ones in blackberries. Environmental factors have little effect, and hence the heritability of pyrene size is high (97%), and phenotypical selection of parents for breeding is efficient (Moore *et al.*, 1975).

Although small pyrenes are preferred, their small size must be associated with good development of the soft tissues which surround them. This quality

Table 15.1. Some correlation coefficients in blackberries between fruit weight and pyrene variables. (A) Data of Moore *et al.* (1974a); (B) Data of Jennings and Brydon (unpublished).

	Inflorescence position		
Characters correlated	Primary (B)	Tertiary (A)	Quaternary (B)
Fruit wt. *vs* mean pyrene wt.	0.48	0.51	0.27
Fruit wt. *vs* pyrene no. (=drupelet no.)	0.67	0.56	—
Fruit wt. *vs* pyrene wt. per fruit	0.83	0.81	0.83
Fruit wt. *vs* fruit wt. per pyrene	0.72	0.35	—
Pyrene wt. *vs* fruit wt. per pyrene	0.69	0.64	—

is assessed by dividing the mean fruit weight by the fruits' mean number of pyrenes, to give an estimate of fruit weight per pyrene. Unfortunately, small pyrenes tend to be associated with small drupelets, i.e. a low fruit weight per pyrene, and this usually gives small fruits, though fruit weight depends upon both the number of drupelets present and their mean weight (Moore *et al.*, 1974a). Some relationships are given in Table 15.1.

15.2 SEED STRUCTURE AND DEVELOPMENT

A raspberry or blackberry carpel has an ovary with two ovules, of which the upper one usually aborts after differentiation of the lower one into a megaspore mother cell. Both of the ovules reach maturity only rarely in diploid raspberries, more frequently in tetraploid raspberries and regularly in some blackberries. The ovules are anatropus, crassinucellate and have a single integument of approximately six layers of cells; the two outer ones resist crushing and can be identified in mature seeds as part of the testa. The inner epidermis facing the nucellus forms an osmotic barrier and seems to have some kind of specialized function. Megaspore formation is normal and the embryo sac is eight-nucleate and usually differentiated at anthesis, though exceptions occur, especially in blackberries, where every stage from megaspore to mature megagametophyte can be found in one flower (Topham, 1970).

Fertilization usually occurs on the day after pollination and the endosperm nucleus begins to divide a day later. In raspberries it remains free-nuclear until after the eighth day, when as many as 400 nuclei may be present (Topham, 1970). In blackberries cell wall formation begins between the tenth and fifteenth day, starting at the micropylar end and progressing towards the chalazal end (Kerr, 1954). It forms a densely packed tissue of approximately cubic cells, which are usually two or three layers thick, though it is thicker around the radicle and along the edges of the cotyledons because it fills all the space between the embryo and the testa. The root tip is not covered. The endosperm is interrupted in the chalazal region by a pedestal-like hypostase, derived from the nucellus, which projects slightly into the embryo-sac cavity and may have a secretory function. Apart from this structure the nucellus is a thin-walled unspecialized tissue from which the endosperm derives its nourishment, partly at least by digesting the cells in immediate contact with it (Topham, 1970).

In contrast to the endosperm, the egg cell in the raspberry does not begin to divide until the fourth day after pollination, or even later, and then develops very slowly during the first stage of fruit development. Its nutrients come directly from the cytoplasm of the endosperm, with which it is in direct contact, but there is little synchronization in the development of the two tissues and both tend to show a considerable range of growth by the eighth day (Table 15.2). The early stages of embryogenesis correspond to the Asterad type as described by Johansen, and the later ones follow the Genn variation, like most other *Rosaceous* species. The main embryo growth occurs during the second stage of fruit development, during which it completely fills the endosperm cavity.

During the initial period after fertilization the whole system behaves as a unit, and the embryo, endosperm, testa and endocarp grow alongside each other (Topham, 1970). But fruit growth, once begun, is largely independent of that of the seed within, and well-set fruit are not necessarily fully fertile in terms of seed production. Moreover, the seed is part endosperm and part embryo, and so it is not surprising that a variety of genetic factors affect its size and shape. Seed size is usually, but not always, closely related to endosperm size (coefficients of correlation of 0.82 and 0.88 have been obtained) and so maternal factors influence it most, though Jennings (1971b) found important interactions between maternal and paternal contributions to the seed, apparently involving the same interacting factors as those which affect the endocarp. These interactions had more influence on

Table 15.2 Numbers of embryo cells and endosperm nuclei in two diploid raspberries (Topham, 1970).

	Endosperm at 4 days		Embryo at 4 days		Endosperm at 8 days		Embryo at 8 days	
Genotype	Mean	Range	Mean	Range	Mean	Range	Mean	Range
Malling Jewel	14.7	4–37	2.2	1–5	163.2	124–184	55.2	14–82
Malling 69139	2.5	1–4	1.0	1–2	101.4	35–233	10.9	2–19

seed size than the genetic interactions which led to inbreeding depression of the embryo. This probably explains why there was only a small reduction in seed size attributable to inbreeding. Endosperm size was so little affected by inbreeding that small inbred embryos tended to be surrounded by a relatively large endosperm in a seed of similar size to that of cross-bred ones.

15.3 SEED GERMINATION

Raspberry and blackberry seeds pass through a dormancy phase before they can germinate. In nature, dormancy is broken by the exposure of moist seed to low winter temperatures, often after it has passed through the digestive tract of a bird. The dormancy is caused by the presence of an acidic, ether-soluble growth-inhibiting substance, or substances, whose concentration is highest in the endosperm and progressively lower in the testa and seed: it is probably formed in the endosperm and diffuses into the other tissues. It disappears during the storage of moist seed at 3 ± 2°C, and its disappearance is correlated with the breaking of dormancy. However, a growth-promoting substance or substances are probably formed during this treatment, and again the endosperm has been implicated because an attempt to germinate excised blackberry seeds was not successful unless the seeds were in contact with the endosperm. Other changes which occur during moist-chilling are a decrease in starch, an increase in sugars and an increase in the activity of such enzymes as catalase, peroxidase and lipase (Lasheen and Blackhurst, 1956).

The changes which lead to loss of dormancy can be accelerated by treatment with chemicals, which probably act in various ways, such as eroding and softening the endocarp, thereby facilitating gaseous exchange, leaching away a dormancy factor and acting as oxidizing agents during chemical reactions. A successful pretreatment of raspberry pyrenes, for application prior to six weeks of moist-chilling, consists of twenty minutes immersion in concentrated sulphuric acid, followed by six days of treatment with one per cent calcium hypochlorite with an excess of calcium hydroxide added and the solution changed after three days; the same concentration of thiourea with added calcium hydroxide is also effective. After this treatment, small additional benefits can be obtained by treatment with 500 p.p.m. gibberellic acid prior to sowing, or by providing supplementary light to give a 16 hour photoperiod after treatment (Jennings and Tulloch, 1965).

Success in germinating blackberry seeds has been obtained by similar methods to those used for raspberries (Wenzel and Smith, 1975), but it is more difficult to germinate blackberry seed than raspberry seed, and low germination percentages are often obtained. However, Moore *et al.* (1974b) obtained a mean germination of 17% and a range up to 67% using ice baths to prevent the temperature from reaching injurious levels, and prolonging the pretreatment with concentrated sulphuric acid to three hours prior to four months of moist-chilling. This treatment was considerably more successful than a shorter acid treatment, apparently because endocarp thickness was reduced on average from 2.82 to 2.27 mm; but the treatment is too drastic for small-seeded genotypes. Nybom (1980) successfully treated blackberry seeds for two hours with concentrated sulphuric acid without using ice baths, and obtained the best results when this was followed by the sequence of treatments used by Jennings and Tulloch (1965). Total removal of the endocarp with needles prior to moist-chilling is probably the ideal. Very high germination percentages have been obtained by this method (Kerr, 1954), though Nesme (1985) found that merely nicking the seed was effective provided that the testa and endosperm were injured. Galletta *et al.* (1986) found that excised embryos raised in tissue culture developed better and earlier than those of seeds given routine acid scarification and chilling, especially poorly developed inbred embryos.

Another reason for the poor germination of blackberry seeds is the high proportion of them whose contents are grossly defective. The pyrenes may appear normal, but the seed within them may be totally collapsed or plump at the chalazal end and collapsed at the micropylar end; even enlarged seeds may lack an embryo. One reason for this is the presence of embryos with aneuploid chromosome numbers—consequences of meiotic irregularity (Kerr, 1954)—and another is the occurrence of incompatibility between the several tissues of the seed (Jennings, 1974).

Seeds vary considerably in their capacity to emerge and in the time of their emergence after sowing. In an experiment where dormancy was the main factor limiting seedling emergence, Jennings (1971c) found that variation in percentage emergence and in emergence time depended most on the identity of the maternal parent and was inversely correlated with the seeds' endosperm size, suggesting that dormancy-inducing substances were present in the endosperm in amounts related to its size. The correlation with endocarp size was lower. But where dormancy had been lost more completely, variation in emergence was associated with variation in embryo size caused by inbreeding effects, which greatly reduced embryo size and consequently had adverse effects on both the emergence percentage and emergence time of the seeds. The major gene *S*, which determines spine production, has an associated effect on endosperm size which apparently affects both percentage emergence and emergence time. It is therefore prone to aberrant segregation ratios which vary according to the percentage emergence attained (Jennings, 1972).

In nature, if cross-bred embryos are larger and emerge first they will establish a selective advantage over later emerging, inbred ones; heterozygotes will tend to be selected and heterozygous populations maintained. These variations in seed structure therefore have far-reaching consequences for the breeding system. The time of seed germination is also an adaptive feature; for example, in Sweden, the seed of blackberries from northerly populations and from zones with severe climates tend to germinate earliest (Nybom, 1980).

Appendices

APPENDIX 1

Origin of some recently released cultivars of raspberries, blackberries and *Rubus* hybrids

A. Red raspberries

Cultivar	Origin
Amber	Taylor × Cuthbert. Geneva, New York, 1950
Amity	(Fallred × OR-US 1347) × (Malling 791/45 × Heritage). Corvallis, Oregon, 1984
Antietam	Marcy × Sunrise. College Park, Maryland, 1953
Augustred	Complex. Durham, New Hampshire, 1973
Autumn Bliss	Complex. East Malling, England, 1983
Avon	M. Promise × Cuthbert. Kentville, Nova Scotia, 1967
Barnaulskaya	Viking × Usanka. Barnaul, U.S.S.R.
Boyne	Indian Summer × Chief. Morden, Manitoba, 1960
Canby	Viking × Lloyd George. Corvallis, Oregon, 1953
Carnival	Ottawa × Rideau. Ottawa, Ontario, 1955
Cherokee	Hilton × (Taylor × Ranere). Blacksburg, Virginia, 1972
Chilcotin	Sumner × Newburgh. Vancouver, British Columbia, 1977
Chilliwack	(Sumner × Carnival) × Skeena. Vancouver, British Columbia, 1986
Citadel	Mandarin × (Sunrise × Oregon selection). College Park, Maryland, 1966
Comet	Ottawa × Modawaski. Ottawa, Ontario, 1955
Comox	(Creston × Willamette) × Skeena. Vancouver, British Columbia, 1986
Creston	Unknown. Creston, British Columbia, 1950
Crimson Cone	Latham × Milton. Urbana, Illinois, 1955
Distad	Unknown. Norway, c. 1960
Dormanred	*R. parvifolius* × Dorsett. Mississippi, 1972

Durham	Taylor × Nectar Blackberry (Prob.). Durham, New Hampshire, 1947
Earlibest	Latham open pollinated. Excelsior, Minnesota, 1960
Early Red	Lloyd George × Cuthbert. South Haven, Michigan, 1952
Esenna Pozlata	(Shopska Alena Self) × [(Prussia × Angliiska) × Newburgh]. Bulgaria, 1985
Fairview	(Pynes Royal × Newburgh) × Washington. Corvallis, Oregon, 1961
Fallgold	Complex. Durham, New Hampshire, 1967
Fallred	Complex. Durham, New Hampshire, 1964
Festival	Muskoka × Trent. Ottawa, Ontario, 1971
Fraser	Chief × Viking. Saskatoon, Saskatchewan, 1960
Futura	Pynes Royal × *R. occidentalis*. Arslev, Denmark, 1985
Gardinia	M. Exploit × Rubin. Čačak, Yugoslavia, 1975
Gatineau	Lloyd George × Newman 23. Ottawa, Ontario, 1950
Geloy	Complex. Max-Planck, Germany, 1963
Gevalo	Complex. Max-Planck, Germany, 1963
Glen Clova	Complex. Invergowrie, Scotland, 1969
Glen Isla	Complex. Invergowrie, Scotland, 1974
Glen Moy	Complex. Invergowrie, Scotland, 1982
Glen Prosen	Complex. Invergowrie, Scotland, 1982
Glorya	Pavlovskaya × sib. Leningrad, U.S.S.R.
Granat	Lloyd George × Preussen. Czechoslovakia
Haida	M. Promise × Creston. Vancouver, British Columbia, 1973
Heritage	Durham × (Milton × Cuthbert). Geneva, New York, 1969
Hilton	Newburgh × St. Walfried. Geneva, New York, 1965
Iskia	Unknown. Kostinbrod, Bulgaria
Ithasca	Newburgh selfed. Excelsior, Minnesota, 1956
Jochems Roem	Malling Promise open pollinated. Wageningen, Holland, 1966
Karallovya	Pavlovskaya × sib. Leningrad, U.S.S.R.
Kellaris 5	Faistrup × Preussen. Kvistgaard, Denmark, 1956
Killarney	Indian Summer × Chief. Morden, Manitoba, 1961

Krupna Dvoroda	M. Exploit × Rubin. Čačak, Yugoslavia, 1975
Lake Geneva	Latham × Ranere. Lake Geneva, Wisconsin, 1960
Liberty	Sunrise S_2 × Newburgh. Ames, Iowa, 1976
Lyulin	Newburgh × Bulgarian Rubin. Bulgaria, 1982
Malling Admiral	Complex. East Malling, England, 1971
Malling Delight	Complex. East Malling, England, 1973
Malling Exploit	Newburgh × (Pynes Royal Self × Lloyd George Self F_2). East Malling, England, 1950
Malling Enterprise	Preussen × (Pynes Royal Self × Lloyd George Self). East Malling, England, 1943
Malling Jewel	Preussen × (Pynes Royal Self × Lloyd George Self). East Malling, England, 1950
Malling Joy	Complex. East Malling, England, 1980
Malling Landmark	Preussen × Baumforth A. East Malling, England, 1947
Malling M	Preussen × Lloyd George. East Malling, England
Malling Leo	Complex. East Malling, England, 1975
Malling Orion	Complex. East Malling, England, 1970
Malling Promise	Newburgh × (Pynes Royal Self × Lloyd George Self F_2). East Malling, England, 1944
Madawaska	Lloyd George × Newman 23. Ottawa, Ontario, 1943
Mandarin	(*R. parvifolius* × Taylor) × Newburgh. Raleigh, North Carolina, 1955
Matsqui	Sumner × Carnival. Vancouver, British Columbia, 1969
Meeker	Willamette × Cuthbert. Puyallup, Washington, 1967
Muskoka	Newman 23 × Herbert. Ottawa, Ontario, 1950
Nootka	Carnival × Willamette. Vancouver, British Columbia, 1977
Norna	Preussen × Lloyd George. Njos, Norway, 1961
Nova	Southland × Boyne. Kentville, Nova Scotia, 1981
Orbita	Novost Kuzmina × Korallovaya. U.S.S.R., 1982
Pathfinder	August Red × *R. strigosus*. Cheyenne, Wyoming, 1976
Paul Camazind	Preussen × Lloyd George. Wädenswil, Switzerland, 1955
Pavlovskaya	Novost Kuzmina × Lloyd George. Leningrad, U.S.S.R.

Pocahontas	Hilton × (Taylor × Ranere). Blacksburg, Virginia, 1972
Prestige	S_2 of Taylor × *R. pungens oldhami*. Durham, New Hampshire 1979
Promiloy	Malling Promise × Complex seedling. Max-Planck, Germany, 1963
Puyallup	Washington × Taylor. Puyallup, Washington, 1953
Rakyeta	Pavolvskaya × sib. Leningrad, U.S.S.R.
Reveille	(Indian Summer × Sunrise) × September. College Park, Maryland, 1966
Romy	Lloyd George open-pollinated. Geneva, Switzerland, 1954
Rote Wädenswiler	Preussen × Superlative. Wadenswil, Switzerland, 1960
Rubin	Lloyd George × Preussen. Max-Planck, Germany, 1954
Rucami	Clone 4a × Paul Camenzind. Max-Planck, Germany, 1980
Rumilo	Clone 4a × Promiloy. Max-Planck, Germany, 1980
Rutrago	Clone 4a × Tragilo. Max-Planck, Germany, 1980
Sensation	Indian Summer Selfed. Champaign, Illinois, 1974
Scepter	September × Durham. College Park, Maryland, 1966
Schonemann	Lloyd George × Preussen. Max-Planck, Germany
Sentinel	Sunrise × Milton. College Park, Maryland, 1966
Sentry	Sunrise × Taylor. College Park, Maryland, 1966
September	Marcy × Ranere. Geneva, New York, 1947
Shopska Alena	Unknown. Kostinbrod. Bulgaria
Sirius	Willamette × Schonemann. Holland, 1973
Skeena	Creston × SHRI 6010/52. Vancouver, British Columbia, 1977
Southland	Complex (2nd backcross of *R. parvifolius*). Raleigh, Carbondale, Illinois, 1968
Spica	Malling Jewel × Willamette, Holland, 1973
Sumner	Washington × Tahoma. Pullayup, Washington, 1956
Sygna	Asker × Lloyd George. Njos, Norway, 1961
Tennessee Autumn	Tenn. 181 × Lloyd George. Knoxville, Tennessee, 1948

Tennessee Luscious	Lloyd George × (Van Fleet × Viking). Tennessee, 1944
Tennessee Prolific	Lloyd George × (Van Fleet × Viking). Tennessee, 1948
Tragilo	Complex. Max Planck, Germany, 1963
Trent	Newman 23 × Lloyd George. Ottawa, Ontario, 1943
Thames	Lloyd George × Newman 23. Ottawa, Ontario, 1952
Titan	Hilton × (Newburgh × St Walfried). Geneva, New York, 1986
Trailblazer	August Red × *R. strigosus*. Cheyenne, Wyoming, 1976
Vega (USSR)	Yunost × Novost Kuzmina. U.S.S.R., 1982
Vega (Denmark)	Pyne's Royal × (Pyne's Royal × 140/25). Arsleve, Denmark, 1985
Vetan	Asker × Lloyd George. Njos, Norway, 1961
Washington	Cuthbert × Lloyd George. Puyallup, Washington, 1943
Willamette	Newburgh × Lloyd George. Corvallis, Oregon, 1942
Yunost	Pavlovskaya × Preussen. Leningrad, U.S.S.R.
Zenith	(140/25 × Pyne's Royal) × Pyne's Royal. Arsleve, Denmark, 1985
Zeva 1	Willamette × Paul Camazind. Wädenswil, Switzerland, 1960
Zeva 2	Rote Wadenswiler Willamette. Wädenswil, Switzerland, 1960
Zeva Herbsternte	(Indian Summer × Romy) × Romy. Wädenswil, Switzerland, 1963

B. Black and Purple Raspberries

Allegany	(Manteo Selfed) × Dundee. College Park, Maryland, 1970
Allen	Bristol × Cumberland. Geneva, New York, 1957
Amethyst	Robertson × Cuthbert. Ames, Iowa, 1968
Black Hawk	Quillan × Black Pearl. Ames, Iowa, 1955
Black Knight	Johnson Everbearing Selfed. Champaign, Illinois, 1973
Brandywine	NY631 × Hilton. Geneva, New York, 1976

Clyde	Bristol × (Newburgh × Indian Summer). Geneva, New York, 1961
Ebonee	Cumberland open pollinated. Ames, Iowa, 1962
Huron	Rachel × Dundee. Geneva, New York, 1965
Jewel	(Bristol × Dundee) × Dundee. Geneva, New York, 1973
Lowden	Bristol × Sodus. Ontario, Canada, 1961
Purple Autumn	Bristol × Indian Summer. Urbana, Illinois, 1953
Redman	Purple Raspberry × (Viking × Wild raspberry). Parkside Saskatchewan, Canada
Royalty	(Cumberland × Newburgh) × (Newburgh × Indian Summer). Geneva, New York, 1982
Success	Morrison × Newhampshire 100. Durham, New Hampshire, 1956

C. Arctic Raspberries and their Hybrids

Anna	*R. stellatus* × *R. arcticus*. Umeå, Sweden, 1980
Beata	*R. stellatus* × *R. arcticus*. Umeå, Sweden, 1982
Heija	Merva × M. Promise. Piikkio, Finland, 1975
Heisa	Merva × M. Promise. Piikkio, Finland, 1976
Linda	*R. stellatus* × *R. arcticus*. Umeå, Sweden
Merva	*R. idaeus* × *R. arcticus* F_3 Piikkio, Finland
Mesma	*R. arcticus*, natural strain. Piikkio, Finland, 1972
Mespi	*R. arcticus*, natural strain. Piikkio, Finland, 1972
Sofia	*R. stellatus* × *R. arcticus*. Umeå, Sweden, 1982
Valentina	*R. stellatus* × *R. arcticus*. Umeå, Sweden, 1985

D. Blackberries

Aurora	US-OR616 × US-OR73. Corvallis, Oregon, 1961
Bailey	Unknown. Geneva, New York, 1950
Black Satin	(US 1482 × Darrow) × Thornfree. Carbondale, Illinois, 1974
Brazos	Lawton × Nessberry. College Station, Texas, 1959
Brison	(F_2 of Brainerd × Brazos) × Brazos. College Station, Texas 1977
Carolina	Austin Thornless × Lucretia. Raleigh, North Carolina, 1955
Cherokee	Brazos × Darrow. Fayetteville, Arkansas, 1974
Chester Thornless	SIUS 47 × Thornfree. Carbondale, Illinois, 1985

Cheyenne	Darrow × Brazos. Fayetteville, Arkansas, 1976
Comanche	Brazos × Darrow. Fayetteville, Arkansas, 1974
Darrow	(Eldorado × Brewer) × Hedrick. Geneva, New York, 1958
Dirksen Thornless	(US 1482 × Darrow) × Thornfree. Carbondale, Illinois, 1974
Early June	USDA 266 × North Carolina 36. Experiment, Georgia, 1959
Ebano	Comanche × (Thornfree × Brazos). Fayetteville, Arkansas
Flint	Brainerd × Eldorado. Experiment, Georgia, 1957
Flordagrand	Regal-Ness × *R. trivialis* (F_2). Gainesville, Florida, 1958
Gem	Flint × Early June. Experiment, Georgia, 1967
Georgia Thornless	USDA 1445 × Early June. Experiment, Georgia, 1967
Hedrick	Eldorado × Brewer. Geneva, New York, 1950
Hull Thornless	(US 1482 × Darrow) × Thornfree. Carbondale, Illinois, 1981
Jersey Black	Thornless Evergreen × Eldorado. New Brunswick, New Jersey, 1953
Loch Ness	Complex. Invergowrie, Scotland, 1988
Kotata	(Pacific × Boysen) × (Jenner-1 × Eldorado). Corvallis, Oregon, 1984
Marion	Chehalem × Olallie. Corvallis, Oregon, 1956
Olallie	Black Logan × Young. Corvallis, Oregon, 1950
Oklawaha	Regal Ness × *R. trivialis*. Gainesville, Florida, 1964
Ranger	Dewblack × Eldorado. College Park, Maryland, 1964
Raven	Dewblack × Eldorado. College Park, Maryland, 1962
Regal	Mammoth × Crandall. Sebastopol, California, 1965
Rosborough	(F_2 of Brainerd × Brazos) × Brazos. College Station, Texas, 1977
Shawnee	Cherokee × (Thornfree × Brazos). Fayetteville, Arkansas, 1985
Silvan	US-OR742 × Marion. Melbourne, Victoria
Smoothstem	(Merton Thornless × US 1411) open-pollinated. Beltsville, Maryland, 1966

Thornfree	(Brainerd × Merton Thornless) × (Merton Thornless × Eldorado). Beltsville, Maryland, 1966
Williams	Himalaya × Taylor. Raleigh, North Carolina, 1962
Womack	(F_2 of Brainerd × Brazos) × Brazos. College Station, Texas, 1977

E. Blackberry–Raspberry Hybrids

Lincoln Loganberry	Histogenic manipulation of Loganberry. Christchurch, New Zealand, 1986
Sunberry	*R. ursinus* × (Tetraploid Malling Jewel open-pollinated). East Malling, England, 1981
Fertödi Bötermö	F_3 of Logan × (*R. caesius* × L. George). Fertöd, Hungary 1980
Tayberry	Aurora × Complex Tetraploid Raspberry. Invergowrie, Scotland, 1979
Tummelberry	Tayberry × Sib of Tayberry, Invergowrie. Scotland, 1983

APPENDIX 2

Gene list for *Rubus*

Symbol	*Gene Effect*	Rubus *species*	*Authority*
A_1	Resistance to *Amphorophora idaei* strains 1 and 3	*idaeus*	Knight *et al.* (1959)
A_2	Resistance to *A. idaei* strain 2	*idaeus*	Knight *et al.* (1960)
A_1A_3	Resistance to *A. idaei* strains 1, 2, 3	*idaeus*	Knight *et al.* (1960)
A_3A_4	Resistance to *A. idaei* strain 2	*idaeus*	Knight *et al.* (1960)
A_5	Resistance to *A. idaei* strain 1	*idaeus*	Knight *et al.* (1960)
A_6	Resistance to *A. idaei* strain 1	*idaeus*	Knight *et al.* (1960)
A_7	Resistance to *A. idaei* strain 1	*idaeus*	Knight *et al.* (1960)
A_8	Resistance to *A. idaei* strains 1, 2, 3, 4	*idaeus*	Knight (1962)
A_9	Resistance to *A. idaei* strains 1, 2, 3, 4	*idaeus*	Knight (1962)
A_{10}	Resistance to *A. idaei* strains 1, 2, 3, 4	*occidentalis*	Keep and Knight (1967)
$A_{L503}(=A_{10})$	Resistance to *A. idaei* strains 1, 2, 3, 4	*occidentalis*	Keep *et al.* (1970)
A_{K4a}	Resistance to *A. idaei* strains 1, 2, 3, 4	*idaeus*	Keep *et al.* (1970)
A_{cor1}	Resistance to *A. idaei* strains 1, 2, 3, 4	*coreanus*	Keep *et al.* (1970)
A_{cor2}	Resistance to *A. idaei* strain 2	*coreanus*	Keep *et al.* (1970)
Ag_1	Resistance to *A. agathonica*	*idaeus*	Daubeny (1966)
Ag_2Ag_3	Resistance to *A. agathonica*	*idaeus*	Daubeny and Stary (1982)
An	Pink petals	*coreanus*	Keep *et al.* (1977a)
$AB(?=A_{K4a})$	Resistance to *A. idaei*	*idaeus*	Baumeister (1962)
B	Waxy bloom on stems	*idaeus*	Lewis (1939)

Bd_1Bd_2	Accessory buds	*idaeus*	Keep (1968c)
$B1B1$	Fruit colour: $B1B1$ = black $B1b1$ = purple	*occidentalis*	Britton *et al.* (1959)
Br_1Br_2	Branched canes	*coreanus*	Keep *et al.* (1977a)
Bu	Immunity from bushy dwarf virus (common strain)	*idaeus*	Jones *et al.* (1982)
C	Pigmented growth	*idaeus*	Crane and Lawrence (1931)
cr	Crumbly fruit (semi-fertility)	*idaeus*	Jennings (1967b)
$d(=sx_3)$	Sepaloid	*idaeus*	Lewis (1939); Keep (1964)
d_1d_2	Sturdy dwarf	*idaeus*	Keep (1969a)
d_3d_4	Crumpled dwarf	*idaeus*	Keep (1969a)
$d_5 = fr$	Frilly dwarf	*idaeus*	Knight *et al.* (1959)
$dw(=d_1d_2)$	Dwarf	*idaeus*	Jennings (1967a)
$f(=sx_1)$	Female-sterile, obtuse foliage	*idaeus*	Crane and Lawrence (1931)
g	Pale-green leaf	*idaeus*	Lewis (1939)
H	Pubescent cane	*idaeus*	Crane and Lawrence (1931)
I_{am}	Immunity from arabis mosaic virus	*idaeus*	Jennings (1964b)
I_{rr}	Immunity from raspberry ringspot virus (common strain)	*idaeus*	Jennings (1964b)
I_{tb}	Immunity from tomato black ring virus	*idaeus*	Jennings (1964b)
i	Yellow fruit	*idaeus*	Keep (1984)
L_1	Strong lateral development, especially calyx and fruit	*idaeus*	Jennings (1966a)
l_2	Reduced lateral development with miniature fruit	*idaeus*	Jennings (1966b)
Ls	Symptoms of infection by raspberry leaf spot virus	*idaeus*	Jones and Jennings (1980)
Lm	Symptoms of infection by raspberry leaf mottle virus	*idaeus*	Jones and Jennings (1980)
$m(=sx_2)$	Male-sterile flower	*idaeus*	Crane and Lawrence (1931)
$no(?=d_1d_2)$	Dwarf	*idaeus*	Rietsema (1939)
P	Intensifies pigment in fruit and spines	*idaeus*	Crane and Lawrence (1931)
R	Rhamnoside-containing pigments	*idaeus*	Barritt and Torre (1975)

s	Spine-free canes and eglandular cotyledons	*idaeus and rusticanus*	Lewis (1939)
S^1 to S^5	Allelic series for incompatibility	*idaeus*	Keep (1968a, 1985)
S^f	Self-fertility (no pollen reaction)	*idaeus*	Keep (1968a)
S_{TE}	Spine-free canes	*laciniatus*	Hall *et al.* (1986a)
Sf	Spine-free canes	unknown	Jennings (1984)
Sf_W	Spine-free canes and eglandular cotyledons	*idaeus*	D. L. Jennings (unpublished)
Sf_L	Spine-free canes	*loganobacus*	Rosati *et al.* (1988)
sl	Simple leaf	*idaeus*	Jennings (1967b)
sk_1	Suckering	*idaeus*	Knight and Keep (1960)
sk_2sk_3	Where homozygous, epistatic to Sk_1sk_1	*idaeus*	Knight and Keep (1960)
So	Sophorose-containing pigments	*idaeus*	Jennings and Carmichael (1980)
Sp_1Sp_2	Resistance to *Sphaerotheca macularis*	idaeus	Keep (1968b)
sp_3	Resistance to *sphaerotheca macularis*	idaeus	Keep (1968b)
sx_4	Sterility	*idaeus*	Keep *et al.* (1977c)
t	Yellow fruit, green spines	*idaeus*	Crane and Lawrence (1931)
Tr	Tip-rooting	*occidentalis*	Knight and Keep (1960)
$w(?=S^5)$	Pollen-tube inhibition	*idaeus*	Lewis (1940); Keep (1985)
wh	Lethal affecting $H:h$ segregation	*idaeus*	Jennings (1967a)
ws	Lethal affecting $S:s$ segregation	*idaeus*	Jennings (1967a)
$wt(?=w)$	Lethal affecting $T:t$ segregation	*idaeus*	Jennings (1967a)
Wr	Whirled receptacle	*idaeus*	Jennings (1977)
x	Red hypocotyl	*idaeus*	Lewis (1939)
Xy	Xylose-containing pigments	*occidentalis*	Jennings and Carmichael (1980)
Y	Yellow fruit	*phoenicolasius*	Jennings and Carmichael (1975b)
Ys	Suppresses gene *Y*	*phoenicolasius*	Jennings and Carmichael (1975b)
Yr	Resistance to yellow rust	*idaeus*	Anthony *et al.* (1986)
Y_{cor}	Yellow fruit	*coreanus*	Jennings and Carmichael (1980)

References

Achmet, S., Kollányi, L., Porpáczy, A. and Szilagyi, K. (1980). Procedures for the production of virus-free stocks of small fruits in Hungary. *Acta Horticult.* **95**, 83–85.

Anon. (1938). "Report of the John Innes Horticultural Institute for 1937", pp. 12–15.

Anon. (1945). Hybridization of black raspberries to secure varieties immune to anthracnose. *In* "Report of Iowa Agricultural Experimental Station", Part 1, p. 296.

Anon. (1969). Fruit breeding. *In* "Report of East Malling Research Station for 1968", pp. 22–25.

Anon. (1984). Boysenberry: clonal selections—district trials. *In* "Annual Report (1983–84) Riwaka Research Station", p. 22. New Zealand Department of Scientific and Industrial Research.

Anthony, V. M., Williamson, B., Jennings, D. L. and Shattock, R. C. (1986). Inheritance of resistance to yellow rust (*Phragmidium rubi-idaei*) in red raspberry. *Ann. Appl. Biol.* **109**, 365–374.

Arasu, N. T. (1968). Overcoming self-incompatibility by irradiation. *In* "Report of East Malling Research Station for 1967", pp. 109–112.

Atkinson, D. (1973). Seasonal changes in the length of white unsuberized root on raspberry plants grown under irrigated conditions. *J. Horticult. Sci.* **48**, 413–419.

Bailey, L. H. (1898). "Sketch of the Evolution of our Native Fruits" Macmillan & Co., New York, 472pp.

Baker, E. A., Batt, R. F., Silva Fernandes, A. M. and Martin, J. T. (1964). Cuticular waxes of plant species and varieties. *In* "Annual Report Long Ashton Agriculture and Horticulture Research Station for 1963", pp. 106–110.

Bammi, R. K. (1965a). Cytogenetics of *Rubus*. IV. Pachytene morphology of *Rubus parvifolius* L. chromosome complement. *Can. J. Genet. Cytol.* **7**, 254–258.

Bammi, R. K. (1965b). "Complement fractionation" in a natural hybrid between *Rubus procerus* Muell. and *R. laciniatus* Willd. *Nature* **208**, 608.

Barbara, D. J., Ashby, S. C. and Knight, V. H. (1985). The occurrence and distribution of isolates of raspberry bushy dwarf virus in England. *Ann. Appl. Biol.* **106**, 75–81.

Barrientos, F. P. and Rodriguez, J. A. (1980). Transgressive segregation for winter chilling requirement in the red raspberry cultivar Malling Exploit. *Acta Horticult.* **112**, 21–24.

Barritt, B. H. and Torre, L. C. (1975). Inheritance of fruit anthocyanin pigments of red raspberry cultivars. *J. Am. Soc. Horticult. Sci.* **100**, 526–528.

Barritt, B. H. and Torre, L. C. (1980). Red raspberry breeding in Washington with emphasis on fruit rot resistance. *Acta Horticult.* **112**, 25–30.

Barritt, B. H., Crandall, P. C. and Bristow, P. R. (1979). Breeding for root rot resistance in red raspberry. *J. Am. Soc. Horticult. Sci.* **104**, 92–94.

Barritt, B. H., Torre, L. C., Pepin, H. S. and Daubeny, H. A. (1980). Fruit firmness measurements in red raspberry. *HortScience* **15**, 38–39.

Basak, W. (1974). Yellow spot—a virus disease of raspberry. *Bull. Acad. Polon. Sci. Ser. Sci. Biol.* Cl V **22**, 47–51.

Bassols, M. M. and Moore, J. N. (1981). "Ebano" thornless blackberry. *HortScience* **16**, 686–687.

Bauer, R. (1961). Is it possible to count on varietal improvement in bush fruits which will be of interest in commercial cultivation? *Erwerbsobstbau* **3**, 185–187.

Bauer, R. (1980). Rucami, Rumilo and Rutrago, three new large-fruited varieties with vector resistance. *Erwerbsobstbau* **22**, 154–158.

Baumann, G. (1982). Elimination of a heat-stable raspberry virus by combining heat treatment and meristem culture. *Acta Horticult.* **129**, 11–12.

Baumeister, G. (1961). Studies on the resistance of raspberry varieties to the virus vectors *Amphorophora rubi* (Kalt.) and *Aphis idaei* (v.d. Goot). *Züchter* **31**, 351–357.

Baumeister, G. (1962). Investigations on the genetics of resistance to the vector, *Amphorophora rubi* (Kalt.) in raspberry varieties. *Züchter* **32**, 1–7.

Bautista, D. (1977). Observations on the cultivation of the Andes berry (*Rubus glaucus*) in the Venezuelan Andes. *Agronomia Trop.* **27**, 253–260.

Beakbane, A. B. (1939). Trials of Loganberries, blackberries and hybrid-berries at East Malling. *In* "Report of East Malling Research Station for 1938", pp. 213–217.

Beakbane, A. B. (1941). Studies of cultivated varieties of *Rubus* and their hybrids. II. Description and selection of clonal races of some cultivated blackberries and hybrid berries, including Loganberries. *J. Pomol.* **18**, 368–378.

Blackman, R. L., Eastop, V. F. and Hills, M. (1977). Morphological and cytological separation of *Amphorophora* Buckton (Homoptera: Aphididae) feeding on European raspberry and blackberry (*Rubus* spp.). *Bull. Entomol. Res.* **67**, 285–296.

Blaupied, G. D. (1972). A study of ethylene in apple, red raspberry and cherry. *Plant Physiol.* **49**, 627–630.

Brainerd, E. V. and Peitersen, A. K. (1920). Blackberries of New England —their classification. *Bull. Vermont Agricult. Exper. St.* **217**, 84pp.

Braun, J. W. and Garth, J. K. L. (1984). Intracane yield compensation in the red raspberry. *J. Am. Soc. Horticult. Sci.* **109**, 526–530.

Breakey, E. P. (1963). Biology and control of the raspberry crown borer *Bembecia marginata* (Harris). *Washington Agricult. Exper. St. Techn. Bull.* No. 39, 13pp.

Briggs, J. B. (1965). The distribution, abundance and genetic relationships of four strains of the *Rubus* aphid (*Amphorophora rubi* (Kalt.)) in relation to raspberry breeding. *J. Horticult. Sci.* **40**, 109–117.

Briggs, J. B., Danek, J., Lyth, M. and Keep, E. (1982). Resistance to the raspberry beetle, *Byturus tomentosus*, in *Rubus* species and their hybrid derivatives with *R. idaeus*. *J. Horticult. Sci.* **57**, 73–78.

Britton, D. M. and Hull, J. W. (1956). Mitotic instability in blackberry seedlings from progenies of Boysen and of Young. *J. Hered.* **47**, 205–210.

Britton, D. M. and Hull, J. W. (1957). Mitotic instability in *Rubus*. *J. Hered.* **48**, 11–20.

Britton, D. M., Lawrence, F. J. and Haut, I. C. (1959). The inheritance of apricot fruit colour in raspberries. *Can. J. Genet. Cytol.* **1**, 89–93.

Brodel, C. F. and Schaefers, G. A. (1980). Evidence for antibiosis in red raspberry to *Aphis Rubicola*. *J. Econom. Entomol.* **73**, 647–650.

Brooks, R. M. and Olmo, H. P. (1944). Register of new fruit and nut varieties. *Proc. Am. Soc. Horticult. Sci.* **45**, 467–490.

Brown, S. W. (1943). The origin and nature of variability in the Pacific coast blackberries (*Rubus ursinus* Cham. & Schlecht and *R. lemurum* sp. nov.). *Am. J. Botany* **30**, 686–697.

Brunt, A. A. and Stace-Smith, R. (1976). The occurrence of the black raspberry latent strain of tobacco streak virus in wild and cultivated *Rubus* species in British Columbia. *Acta Horticult.* **66**, 71–76.

Bruzzese, E. and Hasan, S. (1986). The collection and selection in Europe of isolates of *Phragmidium violaceum* (Uredinales) pathogenic to species of European blackberry naturalised in Australia. *Ann. Appl. Biol.* **108**, 527–533.

Bunyard, E. A. (1922). An introductory note on the history and development of the raspberry. *J. Pomol.* **3**, 5–6.

Bunyard, E. A. (1925). "A Handbook of Hardy Fruits More Commonly Grown in Great Britain". John Murray, London. 258pp.

Burchill, R. T. and Beever, D. L. (1975). Seasonal fluctuations in ascospore concentrations of *Didymella applanata* in relation to raspberry spur blight incidence. *Ann. Appl. Biol.* **81**, 299–304.

Burdon, J. N. (1987). "The role of ethylene in fruit and petal abscission in *Rubus idaeus* L. cv. Glen Clova'. Ph.D. Thesis, University of Stirling, Scotland.

Burn, J. H. and Withell, E. R. (1941). A principle in raspberry leaves which relaxes uterine muscle. *The Lancet* **241**, 1–3.

Butler, E. J., and Jones, S. G. (1949). "Plant Pathology". Macmillan, London. 979pp.

Cadman, C. H. (1948). Some impressions of New Zealand raspberry growing. *In* "Report of East Malling Research Station for 1947", pp. 173–177.

Cadman, C. H. (1961). Raspberry viruses and virus diseases in Britain. *Horticult. Res.* **1**, 47–61.

Card, F. W. (1898). "Bush-Fruits". Macmillan, New York and London.

Christen, H. R. (1950). Researches into the embryology of pseudogamous and sexual *Rubus* species. *Ber. Schweizer. botan. Gesellsch. Bern* **60**, 153–198.

Clausen, J., Keck, D. D. and Hiesey, W. M. (1945). Experimental studies on the nature of species. II. Plant evolution through amphiploidy and autoploidy with examples from the Madiinae. *In* "Carnegie Institution of Washington, Publication 564", Washington, D.C.

Converse, R. H. (1966). Diseases of raspberries and erect and trailing blackberries. "Agricultural Research Service U.S. Department of Agriculture Handbook" No. 310.

Converse, R. H. (1977). Rubus virus diseases important in the United States. *HortScience* **12**, 471–476.

Converse, R. H. (1984). Blackberry viruses in the United States. *Hort-Science* **19**, 185–188.

Converse, R. H. (1986). Sterility disorder of "Darrow" blackberry. *Hort-Science* **21**, 1441–1443.

Converse, R. H. and Schwartze, C. D. (1968). A root rot of red raspberry caused by *Phytophthora erythroseptica. Phytopathol.* **58**, 56–59.

Converse, R. H., Clarke, R. G., Oman, P. W. Sr. and Milbrath, G. M. (1982). Witches' broom disease of black raspberry in Oregon. *Plant Disease* **66**, 949–951.

Craig, D. L. (1960). Studies on the cytology and the breeding behaviour of *Rubus canadensis* L. *Can. J. Genet. Cytol.* **2**, 96–102.

Cram, W. T. and Daubeny, H. A. (1982). Responses of black vine weevil adults fed foliage from genotypes of strawberry, red raspberry, and red raspberry–blackberry hybrids. *HortScience* **17**, 771–773.

Cram, W. T. and Neilson, C. L. (1976). Major insect and mite pests of berry crops in B.C. *Br. Columbia Dept. Agricult. Bull.* 75-9.

Crandall, P. C. and Chamberlain, J. D. (1972). Effects of water stress, cane size and growth regulators on floral primordia development in red raspberries. *J. Amer. Soc. Horticult. Sci.* **97**, 418–419.

Crandall, P. C., Allmendinger, D. F., Chamberlain, J. D. and Biderbost, K. A. (1974a). Influence of cane number and diameter, irrigation, and carbohydrate reserves on the fruit number of red raspberries. *J. Am. Soc. Horticult. Sci.* **99**, 524–526.

Crandall, P. C., Chamberlain, J. D. and Biderbost, K. A. (1974b). Cane characteristics associated with berry number of red raspberry. *J. Am. Soc. Horticult. Sci.* **99**, 370–372.

Crane, M. B. (1935). Blackberries, hybrid berries and autumn-fruiting raspberries. *In* "Cherries and Soft Fruits Conference, Royal Horticultural Society of London", pp. 121–128.

Crane, M. B. (1940). Reproductive versatility in *Rubus*. I. Morphology and inheritance. *J. Genet.* **40**, 109–118.

Crane, M. B. (1946). Pomology department. *In* "Report of John Innes Horticultural Institute for 1945", pp. 6–12.

Crane, M. B. and Darlington, C. D. (1927). The origin of new forms in *Rubus*. *Genetica* **9**, 241–278.

Crane, M. B. and Lawrence, W. J. C. (1931). Inheritance of sex, colour and hairiness in the raspberry, *Rubus idaeus* L. *J. Genet.* **24**, 243–255.

Crane, M. B. and Thomas, P. T. (1949). Reproductive versatility in *Rubus*. III. Raspberry–blackberry hybrids. *Heredity* **3**, 99–107.

Czapik, R. (1981). Elementary apomictic processes in *Rubus* L. *Acta Soc. Botan. Polon.* **50**, 201–204.

Dale, A. (1977). Yield responses to cane vigour control. *Bull. Scottish Horticult. Res. Inst. Assoc.* **13**, 12–18.

Dale, A. (1979). Varietal differences in the relationships between some characteristics of red raspberry fruiting laterals and their position on the cane. *J. Horticult. Sci.* **54**, 257–265.

Dale, A. and Ingram, R. (1981). Chromosome numbers of some South American blackberries. *Horticult. Res.* **21**, 107.

Dale, A. and Topham, P. B. (1980). Fruiting structure of the red raspberry: multivariate analysis of lateral characteristics. *J. Horticult. Sci.* **55**, 397–408.

Darrow, G. M. (1929). Thornless sports of the Young dewberry. *J. Hered.* **20**, 567–569.

Darrow, G. M. (1931a). European blackberry seedlings and hybrids in the Pacific Northwest. *J. Hered.* **22**, 143–146.

Darrow, G. M. (1931b). A productive thornless sport of the Evergreen blackberry. *J. Hered.* **22**, 405–406.
Darrow, G. M. (1937). Blackberry and raspberry improvement. *In* "Yearbook of the United States Department of Agriculture", pp. 496–533.
Darrow, G. M. (1955a). Blackberry–raspberry hybrids. *J. Hered.* **46**, 67–71.
Darrow, G. M. (1955b). The giant Colombian blackberry of Ecuador. *Fruit Var. Horticult. Dig.* **10**, 21–22.
Darrow, G. M. and Longley, A. E. (1933). Cytology and breeding of *Rubus macropetalus*, the Logan, and related blackberries. *J. Agricult. Res.* **47**, 315–330.
Darrow, G. M. and Waldo, G. F. (1933). Pseudogamy in blackberry crosses. *J. Hered.* **24**, 313–315.
Daubeny, H. A. (1966). Inheritance of immunity in the red raspberry to the North American strain of the aphid *Amphorophora rubi* Kltb. *Proc. Am. Soc. Horticult. Sci.* **88**, 346–351.
Daubeny, H. A. (1971). Self-fertility in red raspberry cultivars and selections. *J. Am. Soc. Horticult. Sci.* **96**, 588–591.
Daubeny, H. A. (1980). Red raspberry cultivar development in British Columbia with special reference to pest response and germplasm exploitation. *Acta Horticult.* **112**, 59–66.
Daubeny, H. A. and Pepin, H. S. (1975). Assessment of some red raspberry cultivars and selections as parents for resistance to spur blight. *HortScience* **10**, 404–405.
Daubeny, H. A. and Pepin, H. S. (1981). Resistance of red raspberry fruit and canes to Botrytis. *J. Am. Soc. Horticult. Sci.* **106**, 423–426.
Daubeny, H. A. and Stary, D. (1982). Identification of resistance to *Amphorophora agathonica* in the native North American red raspberry. *J. Am. Soc. Horticult. Sci.* **107**, 593–597.
Daubeny, H. A., Crandall, P. C. and Eaton, G. W. (1967). Crumbliness in the red raspberry with special reference to the "Sumner" variety. *Proc. Am. Soc. Horticult. Sci.* **9**, 224–230.
Daubeny, H. A., Freeman, J. A. and Stace-Smith, R. (1975). Effects of tomato ringspot virus on drupelet set of red raspberry cultivars. *Can. J. Plant Sci.* **55**, 755–759.
Daubeny, H. A., Freeman, J. A. and Stace-Smith, R. (1982). Effects of raspberry bushy dwarf virus on yield and cane growth in susceptible red raspberry cultivars. *HortScience* **17**, 645–647.
Daubeny, H. A., Stace-Smith, R. and Freeman, J. A. (1978). The occurrence and some effects of raspberry bushy dwarf virus in red raspberry. *J. Am. Soc. Horticult. Sci.* **103**, 519–522.
Daubeny, H. A., Pepin, H. S. and Barritt, B. H. (1980). Postharvest *Rhizopus* fruit rot resistance in red raspberry. *HortScience* **15**, 35–37.

Dickson, A. T. (1979). "The Population Dynamics of Raspberry Aphids in Eastern Scotland". Ph.D. Thesis, University of Dundee, Scotland.

Dijkstra, J. (1973). Witches broom disease of blackberries. *Fruilteelt* **63**, 1235–1236.

Dorsey, M. J. (1921). Hardiness from the horticultural point of view. *Proc. Am. Soc. Horticult. Sci.* **18**, 173–178.

Doughty, C. C., Crandall, P. C. and Shanks, C. H. Jr. (1972). Cold injury to red raspberries and the effect of premature defoliation and mite damage. *J. Am. Soc. Horticult. Sci.* **97**, 670–673.

Dowrick, G. J. (1961). Biology of reproduction in *Rubus*. *Nature* **191**, 680–682.

Dowrick, G. J. (1966). Breeding systems in tetraploid *Rubus* species. *Genet. Res.* **7**, 245–253.

Drain, B. D. (1939). Red raspberry breeding for southern adaptation. *Proc. Am. Soc. Horticult. Sci.* **36**, 302–304.

Drain, B. D. (1956). Inheritance in black raspberry species. *Proc. Am. Soc. Horticult. Sci.* **68**, 169–170.

Duclus, J. and Latrasse, A. (1971). Comparison of fixed and volatile components of raspberry varieties at different stages of maturity. *Ann. Technol. Agricole* **20**, 141–151.

Duncan, J. M., Kennedy, D. M. and Seemüller, E. (1987). Identities and pathogenicities of *Phytophthora* spp. causing root rot of red raspberry. *Plant Pathol.* **36**, 276–289.

Einset, J. (1947). Chromosome studies in *Rubus*. *Gentes Herbarium* **7**, 181–192.

Einset, J. (1951). Apomixis in American polyploid blackberries. *Am. J. Botany* **38**, 768–772.

Einset, J. and Pratt, C. (1954). Hybrids between blackberries and red raspberries. *Proc. Am. Soc. Horticult. Sci.* **63**, 257–261.

Ellis, M. A. and Ellett, C. W. (1981). Late leaf rust on Heritage red raspberry in Ohio. *Plant Disease* **65**, 924.

Focke, W. O. (1910–1914). Species Ruborum Bibliotheca Botanica **72**, 1–233; **83**, 1–274. E. Schweizerbartsche Verlagsbuchhandlung, Stuttgart.

Frazier, N. W. (ed.) (1970). "Virus Diseases of Small Fruits and Grapevines". University of California, Division Agricultural Sciences, Berkeley, California.

Freeman, J. A. (1965). Effect of fungicide field sprays on postharvest fruit rot of raspberries. *Can. Plant Disease Survey* **45**, 107–110.

Galletta, G. J., Draper, A. D. and Puryear, R. L. (1986). Characterization of *Rubus* progenies from embryo culture and from seed germination. *Acta Horticult.* **183**, 83–86.

Gerard, B. M. (1985). *Tetrastichus halidayi* (Graham) (Hym. *Eulophidae*) reared from the raspberry beetle, *Byturus tomentosus* (Deg.) (Col. *Byturidae*). *Entomol. Mon. Mag.* **121**, 234.

Gerlach, D. (1965). Fertilization and autogamy in *Rubus caesius. Biolog. Zeutralblatt* **84**, 611–633.

Gordon, S. C. and McKinlay, R. G. (1986). Loganberry cane fly—a new pest of *Rubus* in Scotland. *Crop Res.* **26**, 121–126.

Gordon, S. C. and Taylor, C. E. (1976). Some aspects of the biology of the raspberry leaf and bud mite (*Phyllocoptes (Eriophyes) gracilis* Nal.) Eriophyidae in Scotland. *J. Horticult. Sci.* **51**, 501–508.

Gordon, S. C. and Taylor, C. E. (1977). Chemical control of the raspberry leaf and bud mite, *Phyllocoptes gracilis* (Nal.) (*Eriophyidae*). *J. Horticult. Sci.* **52**, 517–523.

Green, A. (1971). Soft Fruits. *In* "The Biochemistry of Fruits and their Products" (ed. A. C. Hulme). Academic Press, London and New York.

Griffin, G. D., Anderson, J. L. and Jorgenson, E. C. (1968). Interaction of *Meloidogyne hapla* and *Agrobacterium tumefaciens* in relation to raspberry cultivars. *Plant Disease Rep.* **52**, 492–493.

Griffith, B. S. and Robertson, W. M. (1984). Nuclear changes induced by the nematode *Xiphinema diversicaudatum* in root tips of strawberry. *Histochem. J.* **16**, 265–273.

Griffith, J. P. (1925). The Queensland raspberry, *Rubus probus*, a species adapted to tropical conditions. *J. Hered.* **16**, 328–334.

Grisebach, H. (1982). Biosynthesis of anthocyanins. In 'Anthocyanins as food colours' (Ed. P. Markakis), Academic Press, New York.

Grubb, N. H. (1922). Commercial raspberries and their classification. *J. Pomol.* **3**, 11–35.

Grubb, N. H. (1931). The cropping of raspberry varieties at East Malling. *In* "Report East Malling Research Station for 1928–30", pp. 32–45.

Grubb, N. H. (1935). Raspberry breeding at East Malling 1922–34. *J. Pomol.* **13**, 108–134.

Grünwald, J. and Seemüller, E. (1979). Destruction of the protective properties of raspberry periderm as the result of the degradation of suberin and cell wall polysaccharides by the raspberry cane midge *Thomasiniana theobaldi* Barnes Dipt., Cecidomyiidae). *Z. Pflanzenkrankheiten Pflanzenschutz* **86**, 305–314.

Gustafsson, A. (1943). The genesis of the European blackberry flora. *Acta Universit. Lunden.* **39**, 1–199.

Hall, H. K., Cohen, D. and Skirvin, R. M. (1986a). The inheritance of thornlessness from tissue-culture-derived "Thornless Evergreen" blackberry. *Euphytica* **54**, 891–898.

Hall, H. K., Quazi, M. H. and Skirvin, R. M. (1986b). Isolation of a pure thornless Loganberry by meristem tip culture. *Euphytica* **35**, 1039–1044.

Hall, H. K., Skirvin, R. M. and Braam, W. F. (1986c). Germplasm release of "Lincoln Logan", a tissue culture derived genetic thornless "Loganberry". *Fruit Var. J.* **40**, 134–135.

Hargreaves, A. J. and Williamson, B. (1978). Effect of machine-harvester wounds and *Leptosphaeria coniothyrium* on yield of red raspberry. *Ann. Appl. Biol.* **89**, 37–40.

Harrison, B. D. and Winslow, R. D. (1961). Laboratory and field studies on the relation of arabis mosaic virus to its nematode vector *Xiphinema diversicaudatum* (Micoletzky). *Ann. Appl. Biol.* **49**, 621–633.

Haskell, G. (1954). The history and genetics of the raspberry. *Discovery* **15**, 241–246.

Haskell, G. (1960). The raspberry wild in Britain. *Watsonia* **4**, 238–255.

Haskell, G. (1961). Genetics and the distribution of British *Rubi*. *Genetica* **32**, 118–133.

Haskell, G. (1962). Genetics. *In* "Report of the Scottish Horticultural Research Institute for 1961–2", pp. 54–64.

Haskell, G. and Tun, N. N. (1961). Developmental sequence of chromosome number in a cytologically unstable *Rubus* hybrid. *Genett. Res.* **2**, 10–24.

Hedrick, U. P. (1925). The small fruits of New York. "Report of New York State Agricultural Experimental Station" No. 33, 614 pp.

Helliar, M. V. and Turner, E. A. (1984). Autumn fruiting raspberry trial. I. *In* "National Fruit Trials, Brogdale Experimental Horticulture Station, 1983 Annual Review", pp. 30–34.

Hellman, E. W., Skirvin, R. M. and Otterbacher, A. G. (1982). Unilateral incompatibility between red and black raspberries. *J. Am. Soc. Horticult. Sci.* **107**, 781–784.

Heslop-Harrison, Y. (1953). Cytological studies in the genus *Rubus* L. I. Chromosome numbers in the British *Rubus* flora. *New Phytologist* **52**, 22–39.

Heydecker, W. and Marston, M. E. (1968). Quantitative studies on the regeneration of raspberries from root cuttings. *Horticult. Res.* **8**, 142–146.

Hiirsalmi, H. and Säkö, J. (1976). The nectar raspberry, *Rubus idaeus* × *Rubus arcticus*—a new cultivated plant. *Acta Horticult.* **60**, 151–157.

Hiirsalmi, H. and Säkö, J. (1980). Hybrids of the Arctic bramble species (*Rubus stellatus* × *R. arcticus*). *Acta Horticult.* **112**, 103–108.

Hildebrand, E. M. (1940). Cane gall of brambles caused by *Phytomonas rubi* n.sp. *J. Agricult. Res.* **61**, 685–696.

Hill, A. R. (1952). Insect pests of cultivated raspberries in Scotland. *Trans. Ninth Int. Congr. Entomol.* **1**, 589–592.

Hill, A. R. (1953). Aphids associated with *Rubus* species in Scotland. *Entomolog. Mon. Mag.* **49**, 298–303.

Hill, R. G. (1958). Fruit development of the red raspberry and its relation to nitrogen treatment. *Ohio Agricult. Exper. St. Res. Bull.*, No. 803.

Hong, S. B., Lee, D. K., Kim, Y. H., Kong, S. D., Oh, S. D. and Kim, J. H. (1971). Characteristics of 7 Korean red raspberry lines (*Rubus crataegifolius*) selected as recommendable in Korea. "Research Report of the Office of Rural Development 14", Suwan, Korea.

Hoover, E., Luby, J. and Bedford, D. (1986). Yield components of primocane-fruiting red raspberries. *Acta Horticult.* **183**, 163–166.

Hudson, J. P. (1947). The story of a raspberry variety. *New Zealand J. Agricult.* **75**, 179–180.

Hudson, J. P. (1959). Effects of environment on *Rubus idaeus* L. I. Morphology and development of the raspberry plant. *J. Horticult. Sci.* **34**, 163–169.

Hull, J. W. (1961). Commercial red raspberries in Arkansas? *Arkansas Farming Res.* **10**, 10.

Hull, J. W. (1968). Sources of thornlessness for breeding in bramble fruits. *Proc. Am. Soc. Horticult. Sci.* **93**, 280–288.

Hull, J. W. (1969). Southland red raspberry—a new fruit crop for the south. *Fruit Var. Horticult. Dig.* **23**, 48.

Hull, J. W. and Britton, D. M. (1958). Development of colchicine-induced and natural polyploid breeding lines in the genus *Rubus* (Tourn.) L. *Maryland Agricult. Exper. St. Bull.* **A-91**, 1–63.

Jennings, D. L. (1962). Some evidence on the influence of the morphology of raspberry canes upon their liability to be attacked by certain fungi. *Horticult. Res.* **1**, 100–111.

Jennings, D. L. (1963a). Some evidence on the genetic structure of present-day raspberries and some possible implications for further breeding. *Euphytica* **12**, 229–243.

Jennings, D. L. (1963b). Preliminary studies on breeding raspberries for resistance to mosaic disease. *Horticult. Res.* **2**, 82–96.

Jennings, D. L. (1964a). Some evidence of population differentiation in *Rubus idaeus* L. *New Phytologist* **63**, 153–157.

Jennings, D. L. (1964b). Studies on the inheritance in the red raspberry of immunities from three nematode-borne viruses. *Genetica* **34**, 152–164.

Jennings, D. L. (1964c). Two further experiments on flower-bud initiation and cane dormancy in the red raspberry (var. "Malling Jewel"). *Horticult. Res.* **4**, 14–21.

Jennings, D. L. (1966a). The manifold effects of genes affecting fruit size and vegetative growth in the raspberry. I. Gene L_1. *New Phytologist* **65**, 176–187.

Jennings, D. L. (1966b). The manifold effects of genes affecting fruit size and vegetative growth in the raspberry. II. Gene l_2. *New Phytologist* **65**, 188–191.

Jennings, D. L. (1967a). Balanced lethals and polymorphism in *Rubus idaeus*. *Heredity* **22**, 465–479.

Jennings, D. L. (1967b). Observations on some instances of partial sterility in red raspberry. *Horticult. Res.* **7**, 116–122.

Jennings, D. L. (1971a). Some genetic factors affecting fruit development in raspberries. *New Phytologist* **70**, 361–370.

Jennings, D. L. (1971b). Some genetic factors affecting the development of endocarp, endosperm and embryo in raspberries. *New Phytologist* **70**, 885–895.

Jennings, D. L. (1971c). Some genetic factors affecting seedling emergence in raspberries. *New Phytologist* **70**, 1103–1110.

Jennings, D. L. (1972). Aberrant segregation of a gene in the raspberry and its association with effects on seed development. *Heredity* **29**, 83–90.

Jennings, D. L. (1974). Aspects of fruit and seed development which affect the breeding behaviour of *Rubus* species. *Genetica* **45**, 1–10.

Jennings, D. L. (1977). Somatic mutation in the raspberry. *Horticult. Res.* **17**, 61–63.

Jennings, D. L. (1979a). Resistance to *Leptosphaeria coniothyrium* in the red raspberry and some related species. *Ann. Appl. Biol.* **93**, 319–326.

Jennings, D. L. (1979b). Genotype–environment relationships for ripening time in blackberries and prospects for breeding an early ripening cultivar for Scotland. *Euphytica* **28**, 747–750.

Jennings, D. L. (1979c). The occurrence of multiple fruiting laterals at single nodes of raspberry canes. *New Phytologist* **82**, 365–374.

Jennings, D. L. (1982a). Resistance to *Didymella applanata* in red raspberry and some related species. *Ann. Appl. Biol.* **101**, 331–337.

Jennings, D. L. (1982b). Further evidence on the effects of gene *H*, which confers hairiness, on resistance to raspberry diseases. *Euphytica* **31**, 953–956.

Jennings, D. L. (1983). Inheritance of resistance to *Botrytis cinerea* and *Didymella applanata* in canes of *Rubus idaeus* and relationships between these resistances. *Euphytica* **32**, 895–901.

Jennings, D. L. (1984). A dominant gene for spinelessness in Rubus, and its use in breeding. *Crop Res.* **24**, 45–50.

Jennings, D. L. (1986). Breeding for spinelessness in blackberries and blackberry–raspberry hybrids: A review. *Acta Horticult.* **183**, 59–66.

Jennings, D. L. (1987a). Soft fruit breeding. "Report of the Scottish Crop Research Institute for 1986". p.91.

Jennings, D. L. (1987b). Some effects of secondary dormancy and correla-

tive inhibition on the development of lateral buds of raspberry canes (*Rubus idaeus* L.). *Crop Res.* **27** (in press).
Jennings, D. L. and Carmichael, E. (1975a). Resistance to grey mould (*Botrytis cinerea* FR) in red raspberry fruits. *Horticult. Res.* **14**, 109–115.
Jennings, D. L. and Carmichael, E. (1975b). A dominant gene for yellow fruit in the raspberry. *Euphytica* **24**, 467–470.
Jennings, D. L. and Carmichael, E. (1975c). Some physiological changes occurring in overwintering raspberry plants in Scotland. *Horticult. Res.* **14**, 103–108.
Jennings, D. L. and Carmichael, E. (1979). Colour changes in frozen blackberries. *Horticult. Res.* **19**, 15–24.
Jennings, D. L. and Carmichael, E. (1980). Anthocyanin variation in the genus *Rubus*. *New Phytologist* **84**, 505–513.
Jennings, D. L. and Cormack, M. R. (1969). Factors affecting the water content and dormancy of overwintering raspberry canes. *Horticult. Res.* **9**, 18–25.
Jennings, D. L. and Dale, A. (1982). Variation in the growth habit of red raspberry with particular reference to cane height and node production. *J. Horticult. Sci.* **57**, 197–204.
Jennings, D. L. and Ingram, R. (1983). Hybrids of *Rubus parviflorus* (Nutt.) with raspberry and blackberry, and the inheritance of spinelessness derived from this species. *Crop Res.* **23**, 95–101.
Jennings, D. L. and Jones, A. T. (1986). Immunity from raspberry vein chlorosis virus in raspberry and its potential for control of the virus through plant breeding. *Ann. Appl. Biol.* **108**, 417–422.
Jennings, D. L. and McGregor, G. R. (1987). Resistance to cane spot (*Elsinoë veneta*) in the red raspberry and its relationship to resistance to yellow rust (*Phragmidium rubi-idaei*). *Euphytica* (in press).
Jennings, D. L. and Topham, P. B. (1971). Some consequences of raspberry pollen dilution for its germination and for fruit development. *New Phytologist* **70**, 371–80.
Jennings, D. L. and Tulloch, B. M. (1965). Studies on factors which promote germination of raspberry seeds. *J. Exper. Botany* **6**, 329–340.
Jennings, D. L., Anderson, M. M. and Wood, C. A. (1964). Observations on a severe occurrence of raspberry cane death in Scotland. *Horticult. Res.* **4**, 65–77.
Jennings, D. L., Craig, D. L. and Topham, P. B. (1967). The role of the male parent in the reproduction of *Rubus*. *Heredity* **22**, 43–55.
Jennings, D. L., Carmichael, E. and Costin, J. J. (1972). Variation in the time of acclimation of raspberry canes in Scotland and Ireland and its significance for hardiness. *Horticult. Res.* **12**, 187–200.
Jennings, D. L., Dale, A. and Carmichael, E. (1976). Raspberry and

blackberry breeding at the Scottish Horticultural Research Institute. *Acta Horticult.* **60**, 129–133.
Jennings, D. L., Dale, A. and Carmichael, E. (1977). Raspberry Breeding. *In* "Report of the Scottish Horticultural Research Institute for 1976", p. 46.
Jennings, D. L., Dale, A. and Carmichael, E. (1978). In "Report of the Scottish Horticultural Research Institute for 1977", p. 53.
Jennings, D. L., McNicol, R. J. and Brydon, E. (1986a). In "Fourth Annual Report Scottish Crop Research Institute for 1985", p. 84.
Jennings, D. L., McGregor, G. R., Wong, J. A. and Young, C. E. (1986b). Bud suppression ("Blind bud") in raspberries. *Acta Horticult.* **183**, 285–290.
Jinno, T. (1958). Cytogenetic and cytoecological studies on some Japanese species of *Rubus*. I. Chromosomes. *Botan. Mag. (Tokyo)* **71**, 15–22.
Jones, A. T. (1979). The effects of black raspberry necrosis and raspberry bushy dwarf viruses in Lloyd George raspberry and their involvement in raspberry bushy dwarf disease. *J. Horticult. Sci.* **54**, 267–272.
Jones, A. T. (1982). Distinctions between three aphid-borne latent viruses of raspberry. *Acta Horticult.* **129**, 41–48.
Jones, A. T. (1985). Cherry leaf roll virus. *Assoc. Appl. Biol. Descr. Plant Viruses*, No. 304, 6pp.
Jones, A. T. (1986a). Advances in the study, detection and control of viruses and virus diseases of *Rubus*, with particular reference to the United Kingdom. *Crop Res.* **26**, 127–171.
Jones, A. T. (1986b). Causes of raspberry veinbanding mosaic diseases. *In* "Report Scottish Crop Research Institute for 1985", pp. 161–162.
Jones, A. T. and Jennings, D. L. (1980). Genetic control of the reactions of raspberry to black raspberry necrosis, raspberry leaf mottle and raspberry leaf spot viruses. *Ann. Appl. Biol.* **96**, 59–65.
Jones, A. T. and Mayo, M. A. (1975). Further properties of black raspberry latent virus, and evidence for its relationship to tobacco streak virus. *Ann. Appl. Biol.* **79**, 297–306.
Jones, A. T. and Wood, G. A. (1979). The virus status of raspberries (*Rubus idaeus* L.) in New Zealand. *New Zealand J. Agricult. Res.* **22**, 173–182.
Jones, A. T., Murant, A. F., Jennings, D. L. and Wood, G. A. (1982). Association of raspberry bushy dwarf virus with raspberry yellows disease; reaction of *Rubus* species and cultivars, and the inheritance of resistance. *Ann. Appl. Biol.* **100**, 135–147.
Jones, A. T., Gordon, S. C. and Jennings, D. L. (1984). A leaf-blotch disorder of Tayberry associated with the leaf and bud mite (*Phyllocoptes gracilis*) and some effects of three aphid-borne viruses. *J. Horticult. Sci.* **59**, 523–528.

Kallio, H. (1976a). Identification of steam-distilled aroma compound in the press juice of arctic bramble, *Rubus arcticus* L. *J. Food Sci.* **41**, 555–562.
Kallio, H. (1976b). Development of volatile aroma compounds in arctic bramble, *Rubus arcticus* L. *J. Food Sci.* **41**, 563–566.
Kattan, A. A., Albritton, G. A., Nelson, G. S. and Benedict, R. H. (1965). Quality of machine-harvested blackberries. *Arkansas Farm Res.* **14**, 13.
Keane, P. J., Kerr, A. and New, P. B. (1970). Crown gall of stone fruit. II. Identification and nomenclature of *Agrobacterium* isolates. *Australian J. Biol. Sci.* **23**, 585–595.
Keep, E. (1961). Autumn-fruiting in raspberries. *J. Horticult. Sci.* **36**, 174–185.
Keep, E. (1964). Sepaloidy in the red raspberry, *R. idaeus* L. *Can. J. Genet. Cytol.* **6**, 52–60.
Keep, E. (1968a). Incompatibility in *Rubus* with special reference to *R. idaeus* L. *Can. J. Genet. Cytol.* **10**, 253–262.
Keep, E. (1968b). Inheritance of resistance to powdery mildew, *Sphaerotheca macularis* (Fr.) Jaczewski in the red raspberry, *Rubus idaeus* L. *Euphytica* **17**, 417–438.
Keep, E. (1968c). The inheritance of accessory buds in *Rubus idaeus* L. *Genetica* **39**, 209–219.
Keep, E. (1969a). Dwarfing in the raspberry, *Rubus idaeus* L. *Euphytica* **18**, 256–276.
Keep, E. (1969b). Accessory buds in the genus *Rubus* with particular reference to *R. idaeus*. *Ann. Botany* **33**, 191–204.
Keep, E. (1972). Variability in the wild raspberry. *New Phytologist* **71**, 915–924.
Keep, E. (1983). Powdery mildews of temperate fruit crops. "Proceedings of the 15th International Genetics Congress". New Delhi, pp. 105–118.
Keep, E. (1984). Inheritance of fruit colour in a wild Russian red raspberry seedling. *Euphytica* **33**, 507–515.
Keep, E. (1985). Heterozygosity for self-incompatibility in Lloyd George red raspberry. *Fruit Var. J.* **39**, 5–7.
Keep, E. and Knight, R. L. (1967). A new gene from *R. occidentalis* L. for resistance to strains 1, 2 and 3 of the rubus aphid *Amphorophora rubi* Kalt. *Euphytica* **16**, 209–214.
Keep, E. and Knight, R. L. (1968). Use of the black raspberry (*Rubus occidentalis* L.) and other *Rubus* species in breeding red raspberries. *In* "Report of East Malling Research Station for 1967", pp. 105–107.
Keep, E., Knight, R. L. and Parker, J. H. (1970). Further data on resistance to the raspberry aphid, *Amphorophora rubi* (Kltb.). In "Report of East Malling Research Station for 1969", pp. 129–131.
Keep, E., Knight, V. H. and Parker, J. H. (1977a). The inheritance of

flower colour and vegetative characters in *Rubus coreanus*. *Euphytica* **26**, 185–192.

Keep, E., Knight, V. H. and Parker, J. H. (1977b). *Rubus coreanus* as donor of resistance to cane diseases and mildew in red raspberry breeding. *Euphytica* **26**, 505–510.

Keep, E., Knight, V. H. and Parker, J. H. (1977c). An association between response to mildew (*Sphaerotheca macularis* (Fr.) Jaczewski), sex and spine colour in the raspberry. *J. Horticult. Sci.* **52**, 193–198.

Keep, E., Parker, J. H. and Knight, V. H. (1980). Recent progress in raspberry breeding at East Malling. *Acta Horticult.* **112**, 117–125.

Keep, E., Parker, J. H. and Knight, V. H. (1982). Malling Sunberry. *In* "Report of East Malling Research Station for 1981", pp. 183–184.

Kennedy, G. C. and Schaefers, G. A. (1975). Role of nutrition in the immunity of red raspberry to *Amphorophora agathonica* Hottes. *Environ. Entomol.* **4**, 115–119.

Kennedy, J. S., Day, M. F. and Eastop, V. L. F. (1962). "A Conspectus of Aphids as Vectors of Plant Viruses". Commonwealth Institute of Entomology, London, 114pp.

Kennedy, G. C., Schaefers, G. A. and Ourecky, D. K. (1973). Resistance in red raspberry to *Amphorophora agathonica* Hottes and *Aphis rubicola* Oesthund. *HortScience* **8**, 311–313.

Kerr, E. A. (1954). Seed development in blackberries. *Can. J. Botany* **32**, 654–672.

Khusnullin, K. K., Kichina, V. V. and Tyurina, M. M. (1984). Winter hardiness of raspberry under controlled conditions. *Sel'skokhozyaistvennaya Biol.* **1**, 62–65.

Kichina, V. V. and Bayanova, L. V. (1979). Inheritance of winter hardiness in the winter progeny of raspberry. *Sb. statei orlov. zonal'n plod.- yagod. opyt. st.* **9**, 69–76.

Kichina, V. V., Khusnullin, K. K. and Tyurina, M. M. (1982). Resistance of raspberry to early frosts. *Dokl. Vs. Ord. Lenina i Ord. Trudovogo Krasnogo Znameni Akad. Sel'skokhozyaistvennykh Nauk Imeni V. I. Lenina* **6**, 23–24.

Kichina, V. V., Gogoleva, G. A. and Barteneva, L. V. (1986). Inheritance of a high degree of winter hardiness in raspberry. *Sadovodstvo* **2**, 25–26.

Knight, R. L. (1962). Heritable resistance to pests and diseases in fruit crops. *In* "Proceedings 16th International Horticultural Congress 1962", Brussels 3, pp. 19–104.

Knight, R. L. and Keep, E. (1960). The genetics of suckering and tip rooting in the raspberry. *In* "Report of East Malling Research Station for 1959", pp. 57–62.

Knight, R. L. and Keep, E. (1966). Breeding new soft fruits. *In*

"Fruit—Present and Future". Royal Horticultural Society, London, pp. 98–111.

Knight, R. L., Keep, E. and Briggs, J. B. (1959). Genetics of resistance to *Amphorophora rubi* (Kalt.) in the raspberry. I. The gene A_1 from Baumforth A. *J. Genet.* **56**, 261–280.

Knight, R. L., Briggs, J. B. and Keep, E. (1960). Genetics of resistance to *Amphorophora rubi* (Kalt.) in the raspberry. II. The genes A_2–A_7 from the American variety, Chief. *Genet. Res.* **1**, 319–331.

Knight, V. H. (1980a). Responses of red raspberry cultivars and selections to *Botrytis cinerea* and other fruit-rotting fungi. *J. Horticult. Sci.* **55**, 363–369.

Knight, V. H. (1980b). Screening for fruit rot resistance in red raspberries at East Malling. *Acta Horticult.* **112**, 127–134.

Knight, V. H. (1986). Recent progress in raspberry breeding at East Malling. *Acta Horticult.* **183**, 67–76.

Kobel, F. and Schütz, F. (1963). Zeva Herbsternte, a new raspberry variety. *Schweizer. Z. Obst Und Weinbau* **72**, 443–445.

Koellreuter, J. (1950). Morphology and biology of *Rhabdospora ramealis* (Desm. et Rob.) Sacc. *Phytopatholog. Z.* **17**, 129–160.

Korotkov, N. I., Kichina, V. V. and Zhukov, O. S. (1985). Wild raspberry—useful material in breeding for winter hardiness. *Sadovodstvo* **2**, 28–29.

Kuminov, E. P. (1956). Perpetual raspberries. *Agrobiologija* No. 5, pp. 148–149.

Kurppa, A. and Martin, R. R. (1986). Use of double-stranded RNA for detection and identification of virus diseases of *Rubus* species. *Acta Horticult.* **186**, 51–62.

Labanowska, B. H. (1978). Studies on the intensity of occurrence of spider mites (Tetranychidae) on several raspberry cultivars. *Prace Inst. Sadownictwa Skierniewicach* **20**, 217–221.

Lamberti, F. and Bleve-Zacheo, T. (1979). Studies on *Xiphinema americanum sensu lato* with descriptions of fifteen new species (*Nematoda, Longidoridae*). *Nematol. Mediterran.* **7**, 51–106.

Larsson, E. G. K. (1969). Experimental taxonomy as a base for breeding in northern *Rubi. Hereditas* **63**, 283–351.

Lasheen, A. M. and Blackhurst, H. T. (1956). Biochemical changes associated with dormancy and after-ripening of blackberry seed. *Proc. Am. Soc. Horticult. Sci.* **67**, 331–340.

Latrasse, A. (1982). Raspberry aroma quality. II. Interpretation of the aroma index. *Lebensmittel-Wissensch. Technol.* **15**, 49–51.

Latrasse, A., Lantin, B., Mussillon, P. and Sanis, J. L. (1982). Raspberry aroma quality. I. Rapid colorimetric determination of an aroma index using vanillin in concentrated sulphuric acid solution. *Lebensmittel-Wissensch. Technol.* **15**, 19–21.

Lawrence, F. J. (1976). Breeding self-supporting fall-cropping red raspberries. *Acta Horticult.* **60**, 145–149.

Lawrence, F. J. (1980). Breeding primocane fruiting red raspberries at Oregon State University. *Acta Horticult.* **112**, 145–149.

Lawrence, F. J. (1986). New thornless blackberries from the Cooperative U.S. Department of Agriculture and Oregon State Experiment Breeding Program. *Proc. Oregon Horticult. Soc.* **77**, 153–154.

Lewis, D. (1939). Genetical studies in cultivated raspberries. I. Inheritance and linkage. *J. Genet.* **38**, 367–379.

Lewis, D. (1940). Genetical studies in cultivated raspberries. II. Selective fertilization. *Genetics* **25**, 278–286.

Lewis, D. (1941). The relationship between polyploidy and fruiting habit in the cultivated raspberry. *In* "Proceedings of the 7th Genetics Congress", Edinburgh, 1939, p. 190.

Lewis, D. and Crowe, L. K. (1958). Unilateral interspecific incompatibility in flowering plants. *Heredity* **12**, 233–256.

Lockshin, L. S. and Elfving, D. C. (1981). Flowering response of "Heritage" red raspberry to temperature and nitrogen. *HortScience* **16**, 527–528.

Longley, A. E. (1924). Cytological studies in the genus *Rubus*. *Am. J. Botany* **11**, 249–282.

Maas, J. L. (1986). Epidemiology of the *Botryosphaeria dothidea* cane canker disease of thornless blackberry. *Acta Horticult.* **183**, 125–130.

Mackenzie, K. A. D. (1979). The structure of the fruit of the red raspberry (*Rubus idaeus* L.) in relation to abscission. *Ann. Botany* **43**, 355–362.

McElroy, F. D. (1975). Nematode control in established raspberry plantations. *In* "Nematode Vectors of Plant Viruses" (eds. F. Lamberte, C. E. Taylor and J. W. Seinhoist), NATO Advanced Study Institute Series, Series A, Life Science Vol. 2, pp. 445–446. Plenum, London.

McElroy, F. D. (1977). Effect of two nematode species on establishment, growth and yield of raspberries. *Plant Disease Rep.* **61**, 277–279.

McGregor, G. R. and Kroon, K. H. (1984). Silvan Blackberry. *HortScience* **19**, 732–733.

McKeen, W. E. (1954). A study of cane and crown galls on Vancouver Island and a comparison of the causal organisms. *Phytopathology* **44**, 651–655.

McNicol, R. J., Williamson, B., Jennings, D. L. and Woodford, J. A. T. (1983). Resistance to raspberry cane midge (*Resseliella theobaldi*) and its association with wound periderm in *Rubus crataegifolius* and its red raspberry derivatives. *Ann. Appl. Biol.* **103**, 489–495.

McNicol, R. J., Williamson, B. and Dolan, A. (1985). Infection of red raspberry styles and carpels by *Botrytis cinerea* and its possible role in post-harvest grey mould. *Ann. Appl. Biol.* **106**, 49–53.

McPheeters, K. and Skirvin, R. M. (1983). Histogenic layer manipulation in chimeral "Thornless Evergreen" trailing blackberry. *Euphytica* **32**, 351–360.

Måge, F. (1975). Dormancy in buds of red raspberry. *Meld. Norges Landbruksh.* **54**, No. 21.

Markarian, D. and Olmo, H. P. (1959). Cytogenetics of Rubus. I. Reproductive behaviour of *Rubus procerus* Muell. *J. Hered.* **50**, 131–136.

Massee, A. M. (1954). "The Pests of Fruits and Hops". 3rd edition. Crosby Lockwood & Son, London. 325pp.

Mason, D. T. (1974). Measurement of fruit ripeness and its relation to mechanical harvesting of the red raspberry (*Rubus idaeus* L.). *Horticult. Res.* **14**, 21–27.

Mason, D. T. and Dennis, C. (1978). Post-harvest spoilage of Scottish raspberries in relation to pre-harvest fungicide sprays. *Horticult. Res.* **18**, 41–53.

Mišić, P. and Tešović, Z. V. (1973). The native red raspberry, *Rubus idaeus* in western Serbia and eastern Montenegro. *Jugoslov. Voćarstva* **7**, 1–9.

Mišić, P., Bugarčič, V. and Tešić, B. (1972). Contribution to the study of inheritance in raspberry. *Jugoslov. Voćarstvo* **4**, 81–89.

Moore, J. N. (1984). Blackberry breeding. *HortScience* **19**, 183–197.

Moore, J. N., Brown, G. R. and Brown, E. D. (1974a). Relationship between fruit size and seed number and size in blackberries. *Fruit Var. J.* **28**, 40–45.

Moore, J. N., Brown, E. D. and Lundergan, C. (1974b). Effect of duration of acid scarification on endocarp thickness and seedling emergence of blackberries. *HortScience* **9**, 204–205.

Moore, J. N., Lundergan, C. and Brown, E. D. (1975). Inheritance of seed size in blackberry. *J. Am. Soc. Horticult. Sci.* **100**, 377–379.

Murant, A. F. (1974). Viruses affecting raspberry in Scotland. *Scottish Horticult. Res. Inst. Assoc. Bull.* **9**, 37–43.

Murant, A. F. and Roberts, I. M. (1971). Mycoplasma-like bodies associated with *Rubus* stunt disease. *Ann. Appl. Biol.* **67**, 389–393.

Murant, A. F. and Roberts, I. M. (1980). Particles of raspberry vein chlorosis virus in the aphid vector, *Aphis idaei. Acta Horticult.* **95**, 31–35.

Murant, A. F., Chambers, J. and Jones, A. T. (1974). Spread of raspberry bushy dwarf virus by pollination, its association with crumbly fruit, and problems of control. *Ann. Appl. Biol.* **77**, 271–281.

Murant, A. F., Jennings, D. L. and Chambers, J. (1973). The problem of crumbly fruit in raspberry nuclear stocks. *Horticult. Res.* **13**, 49–54.

Nesme, X. (1985). Respective effects of endocarp, testa and endosperm, and embryo on the germination of raspberry (*Rubus idaeus* L.). *Can. J. Plant Sci.* **65**, 125–130.

Ness, H. (1925). Breeding experiments with blackberries and raspberries. "Bulletin Texas Agricultural Experiment Station", No. 326, 28pp.

Newton, A. (1980). Progress in British *Rubus* studies. *Watsonia* **13**, 35–40.

Nursten, H. E. (1970). Volatile compounds: the aroma of fruits. *In* "The Biochemistry of Fruits and their Products" (ed. A. C. Hulme), p. 253. Academic Press, London.

Nursten, H. E. and Williams, A. A. (1967). Fruit aromas: a survey of components identified. *Chem. Indust.* **12**, 486–497.

Nybom, H. (1980). Germination in Swedish blackberries (*Rubus* L. subgen. *Rubus*). *Botaniska notiser* **133**, 619–631.

Nybom, H. (1985). Active self-pollination in blackberries (*Rubus* subgen. *Rubus*). *Nordic J. Botany* **5**, 521–525.

Oberle, G. D. and Moore, R. C. (1952). Transmission of the autumn-fruiting character in crosses of red and black raspberries. *Proc. Am. Soc. Horticult. Sci.* **60**, 235–237.

Oberle, G. D., Moore, R. C. and Nicholson, J. O. (1949). Parents used in breeding autumn-fruiting red raspberries for Virginia. *Proc. Am. Soc. Horticult. Sci.* **53**, 269–272.

Oort, A. J. P. (1952). Die-back of blackberry caused by *Septocyta ramealis* (Rob.) Pet. *Tijdsch. Plantenziek.* **58**, 247–250.

Ourecky, D. K. (1975). Brambles. *In* "Advances in Fruit Breeding" (eds. J. Janick and J. N. Moore). Purdue University Press.

Ourecky, D. K. (1976). Fall-bearing red raspberries, their future and potential. *Acta Horticult.* **60**, 135–144.

Ourecky, D. K. (1978). The small fruit breeding programme in New York State. *Fruit Var. J.* **32**, 50–57.

Ourecky, D. K. and Slate, G. L. (1966). Hybrid vigour in *Rubus occidentalis- R. leucodermis* seedlings. *In* "Proceedings of the 17th International Horticultural Congress", Maryland. Abstract 277, Vol. 1.

Overcash, J. P. (1972). Dormanred raspberry: a new variety for Mississippi. "Mississippi State University Bulletin 793", pp. 1–7.

Oydvin, J. (1970). Important parent cultivars in raspberry breeding. "Publ. Statens Forsoksgard Njors", 42pp.

Pascoe, I. G., Washington, W. S. and Guy, G. (1984). White root rot of raspberry in Victoria is caused by a *Vararia* species. *Trans. Br. Mycolog. Soc.* **82**, 723–726.

Pears, L. M. and Davidson, R. H. (1956). "Insect Pests of Farm, Garden and Orchard", 5th edition. J. Wiley/Chapman & Hall, New York and London.

Peitersen, A. K. (1921). Blackberries of New England: Genetic status of the plants. *Bull. Vermont Agricult. Exper. St.* **218**, 34pp.

Pepin, H. S. and MacPherson, E. A. (1980). Some possible factors affecting fruit rot resistance in red raspberry. *Acta Horticult.* **112**, 205–207.

Pepin, H. S., Williamson, B. and Topham P. B. (1985). The influence of cultivar and isolate on the susceptibility of red raspberry canes to *Didymella applanata*. *Ann. Appl. Biol.* **106**, 335–347.

Pepin, H. S., Daubeny, H. A. and Carne, I. C. (1967). Pseudomonas blight of raspberry. *Phytopathology* **57**, 929–931.

Percival, M. S. (1946). Observations on the flowering and nectar secretion of *Rubus fruticosus* (agg). *New Phytologist* **45**, 111–123.

Petkov, B. (1963). A study of the melliferous properties of the cultured raspberry. *Selskostopanska Nauk.* **2**, 201–207.

Pitcher, R. S. (1952). Observations on the raspberry cane midge (*Thomasiniana theobaldi* Barnes). I. Biology. *J. Horticult. Sci.* **27**, 71–94.

Pitcher, R. S. and Webb, P. C. R. (1952). Observations on the raspberry cane midge (*Thomasiniana theobaldi* Barnes). II. "Midge blight", a fungal invasion of the raspberry cane following injury by *T. theobaldi*. *J. Horticult. Sci.* **27**, 95–100.

Polesello, A., Crivelli, G., Geat, S., Senesi, E. and Eccher Zerbini, P. (1986). Research on the quick freezing of berry fruit. II. Colour changes in frozen blackberries. *Crop Res.* **26**, 1–16.

Pool, R. A., Ingram, R., Abbot, R. J., Jennings, D. L. and Topham, P. B. (1981). Karyotype variation in *Rubus* with special reference to *R. idaeus* L. and *R. coreanus* Miquel. *Cytologia* **46**, 125–132.

Pope, W. T. (1930). Report of the Horticultural Division. *In* "Report of Hawaii Agricultural Experimental Station for 1929", pp. 3–22.

Popenoe, W. (1920). The Colombian berry or Giant Blackberry of Colombia. *J. Hered.* **11**, 194–202.

Popenoe, W. (1921). The Andes berry. *J. Hered.* **12**, 386–393.

Pratt, C. and Einset, J. (1955). Development of the embryo sac in some American blackberries. *Am. J. Botany* **42**, 637–645.

Rautapää, J. (1967). Studies on the host-plant relationships of *Aphis idaei* v.d. Goot and *Amphorophora rubi* (Kalt.) (Hom., *Aphididae*). *Ann. Agricult. Fenniae* **6**, 174–190.

Rebandel, Z., Przysiecka, M. and Cofta, H. (1985). The influence of certain fungal diseases on the buds and lateral fruiting shoots of raspberry. *Prace Komisji Nauk Rolniczch Komisji Nauk Lesnych* **59**, 189–201.

Redalen, G. (1977). Fertility in raspberries. *Melding. Norges Landbrukshogskole* **56**, 1–13.

Reeve, R. M. (1954). Fruit histogenesis in *Rubus strigosus*. *Am. J. Botany* **41**, 152–160; 173–181.

Reeve, R. M., Wolford, E. and Nimmo, C. C. (1965). A review of fruit structure and the processing of raspberries. *Food Technol.* **19**, 78–82.

Rietsema, I. (1939). A solution of the mosaic problem in raspberries. *Landbouwkundig Tijdsch.* **57**, 14–25.

Roach, F. A. (1985). "Cultivated Fruits of Britain: Their Origin and History". Blackwell, Oxford.

Robertson, M. (1957). Further investigations of flower-bud development in the genus *Rubus*. *J. Horticult. Sci.* **32**, 265–273.

Rock, J. F. (1921). The Akala berry of Hawaii. Asa Gray's *Rubus macraei*, an endemic Hawaiian raspberry. *J. Hered.* **12**, 146–150.

Rosati, P., Gaggioli, D. and Giunchi, L. (1986). Genetic stability of micropropagated Loganberry plants. *J. Horticult. Sci.* **61**, 33–41.

Rosati, P., Hall, H. K., Jennings, D. L. and Gaggioli, D. (1988). A dominant gene for thornlessness obtained from the thornless Loganberry. *Jn. Amer. Soc. Horticult. Sci.* (in press).

Rousi, A. (1965). Variation among populations of *Rubus idaeus* in Finland. *Ann. Agricult. Fenniae* **41**, 49–58.

Rozanova, M. A. (1934). Modes of form genesis in the genus *Rubus*. *J. Botany, U.R.S.S.* **19**, 376–384.

Rozanova, M. A. (1938). On the polymorphic type of species' origin. *C.R. (Dokl.) Acad. Sci. U.R.S.S.* **18**, 677–680.

Rozanova, M. A. (1939a). Evolution of cultivated raspberry. *C.R. (Dokl.) Acad. Sci. U.R.S.S.* **24**, 179–181.

Rozanova, M. A. (1939b). Role of autopolyploidy in the origin of Siberian raspberry. *C.R. (Dokl.) Acad. Sci. U.R.S.S.* **24**, 58–60.

Rozanova, M. A. (1945). On some species, subspecies and varieties within the conspecies *Rubus idaeus* L. *J. Botany U.R.S.S.* **30**, 44–48.

Satomi, N. and Naruhashi, N. (1971). *Ann. Rep. Botan. Garden, Fac. Sci., Univ. Kanazawa* **4**, 1–17.

Schaefers, G. A., Labanowska, B. H. and Brodel, C. F. (1978). Field evaluation of eastern raspberry fruitworm damage to varieties of red raspberry. *J. Econ. Entomol.* **71**, 566–569.

Scott, D. H. and Ink, D. P. (1966). Origination of "Smoothstem" and "Thornfree" blackberry varieties. *Fruit Var. Horticult. Dig.* **20**, 31–33.

Seemüller, E., Duncan, J. M., Kennedy, D. M. and Riedl, M. (1986). *Phytophthora* sp. als Ursache einer Wurzelfäule an Himbeere. *Nachrichtenblatt Deutsch. Pflanzensch.* **38**, 17–21.

Sherman, W. B. and Sharpe, R. H. (1971). Breeding *Rubus* for warm climates. *HortScience* **6**, 147–149.

Slate, G. L. (1934). The best parents in purple raspberry breeding. *Proc. Am. Soc. Horticult. Sci.* **30**, 108–112.

Slate, G. L. (1938). New or noteworthy fruits. XII. Small fruits. "Bulletin of New York State Agricultural Experimental Station", No. 680, 18pp.
Slate, G. L. (1954). New blackberry varieties. "Bulletin of New York State Agricultural Experimental Station", No. 764, 4pp.
Slate, G. L. (1958). Darrow—a promising new blackberry. *Farming Res.* **24**, 7.
Slate, G. L. and Klein, L. G. (1952). Black raspberry breeding. *Proc. Am. Soc. Horticult. Sci.* **59**, 266–268.
Slate, G. L. and Watson, J. (1964). Progress in breeding autumn-fruiting raspberries. *Farming Res.* **30**, 6–7.
Snir, I. (1983). Chemical dormancy breaking of red raspberry. *HortScience* **18**, 710–713.
Snir, I. (1986). Growing raspberries under subtropical conditions. *Acta Horticult.* **183**, 183–186.
Sobczykiewicz, D. (1986). Elimination of heat-stable raspberry vein chlorosis virus by meristem culture. *Acta Horticult.* **186**, 47–49.
Spiler, E. E. (1948). Winter hardiness and drought resistance of raspberry and dewberry varieties under the conditions of the southern Ukranian S.S.R. *Sad Ogorod* **9**, 36–38.
Stace-Smith, R. (1955). Studies on *Rubus* virus diseases in British Columbia. II. Black raspberry necrosis. *Can. J. Botany* **33**, 314–322.
Stephens, S. G. (1942). Colchicine-produced polyploids in Gossypium. I. An autotetraploid Asiatic cotton and certain of its hybrids with wild diploid species. *J. Genet.* **44**, 272–295.
Swait, A. A. J. (1980). Field observations on disease susceptibility, yield and agronomic characters of some new raspberry cultivars. *J. Horticult. Sci.* **55**, 133–137.
Tammisola, J. and Ryxjnänen, A. (1970). Incompatibility in *Rubus arcticus* L. *Hereditas* **66**, 269–278.
Tate, K. G. (1981). Aetiology of dryberry disease of Boysenberry in New Zealand. *New Zealand J. Exper. Agricult.* **9**, 371–376.
Taylor, C. E. (1971). The raspberry beetle (*Byturus tomentosus*) and its control with alternative chemicals to DDT. *Horticult. Res.* **11**, 107–112.
Taylor, C. E. and Brown, D. J. F. (1976). The geographical distribution of *Xiphenema* and *Longidorus* nematodes in the British Isles and Ireland. *Ann. Appl. Biol.* **84**, 383–402.
Taylor, C. E. and Brown, D. J. F. (1981). Nematode-virus interactions. *In* "Plant Parasitic Nematodes" (eds. B. M. Zuckerman and R. A. Rohde). Vol. 3, pp. 281–301. Academic Press, New York.
Taylor, C. E. and Gordon, S. C. (1975). Further observations on the biology and control of the raspberry beetle (*Byturus tomentosus* (Deg.)) in eastern Scotland. *J. Horticult. Sci.* **50**, 105–112.

Thomas, P. T. (1940*a*). The origin of new forms in *Rubus*. III. The chromosome constitution of *R. loganobaccus* Bailey, its parents and derivatives. *J. Genet.* **40**, 141–156.

Thomas, P. T. (1940*b*). Reproductive versatility in *Rubus*. II. The chromosomes and development. *J. Genet.* **40**, 119–128.

Thompson, M. M. (1961). Cytogenetics of *Rubus*. II. Cytological studies of the varieties Young, Boysen and related forms. *Am. J. Botany* **48**, 667–673.

Thompson, M. M. (1962). Cytogenetics of *Rubus*. III. Meiotic instability in some higher polyploids. *Am. J. Botany* **49**, 575–582.

Topham, P. B. (1967). Fertility in crosses involving diploid and autotetraploid raspberries. I. The embryo. *Ann. Botany* **31**, 673–686.

Topham, P. B. (1970). The histology of seed development in diploid and tetraploid raspberries (*Rubus idaeus* L.). *Ann. Botany* **34**, 123–135.

Trudgill, D. L. (1983). The effect of nitrogen and of controlling *Pratylenchus penetrans* with nematicides on the growth of raspberry (*Rubus idaeus* L.). *Crop Res.* **23**, 103–112.

Trudgill, D. L. and Alphey, T. J. W. (1976). Chemical control of the virus-vector nematode *Longidorus elongatus* and of *Pratylenchus crenatus* in raspberry plantations. *Plant Pathol.* **25**, 15–20.

Ure, C. R. (1963). A study of the parental value of nine red raspberry varieties with respect to combining ability and inheritance of vigour and hardiness. *Dissert. Abstr.* **23**, 2671.

Vaarama, A. (1939). Cytological studies on some Finnish species and hybrids of the genus *Rubus* L. *J. Sci. Agricult. Soc. Finland* **11**, 72–85.

Vaarama, A. (1949). Cytogenetic studies on two *Rubus arcticus* hybrids. *J. Sci. Agricult. Soc. Finland* **20**, 67–69.

Vasilakakis, M. D., McCown, B. H. and Dana, M. N. (1979a). Hormonal changes associated with growth and development of red raspberries. *Physiol. Plantarum* **45**, 17–22.

Vasilakakis, M. D., Struckmeyer, B. E. and Dana, M. N. (1979b). Temperature and development of red raspberry flower buds. *J. Am. Soc. Horticult. Sci.* **104**, 61–62.

Vasilakakis, M. D., McCown, B. H. and Dana, M. N. (1980). Low temperature and flowering of primocane-fruiting red raspberries. *HortScience* **15**, 750–751.

Van Adrichem, M. C. J. (1972). Variation among British Columbia and northern Alberta populations of raspberries, *Rubus idaeus* subspp. *strigosus* Michx. *Can. J. Plant Sci.* **52**, 1067–1072.

Van Broekhoven, L. W., Minderhoud, L., Holland, G. J. J., Tansk, R. J. M. and Lousberg, R. J. H. C. (1975). Purification and properties of a phytotoxic glycopeptide from *Didymella applanata* (Niessl) Sacc. *Phytopathol. Z.* **83**, 49–56.

Vrain, T. C. and Daubeny, H. A. (1986). Relative resistance of red raspberry and related genotypes to the root lesion nematode. *HortScience* **21**, 1435–1437.

Waister, P. D. and Gill, P. A. (1979). Plant hardiness and its significance for the raspberry crop in Scotland. *Sci. Horticult.* **30**, 59–68.

Waldo, G. F. (1934). Fruit bud formation in brambles. *Proc. Am. Soc. Horticult. Sci.* **30**, 263–267.

Waldo, G. F. (1968). Blackberry breeding involving native Pacific coast parentage. *Fruit Var. Horticult. Dig.* **22**, 3–7.

Waldo, G. F. and Darrow, G. M. (1941). Breeding autumn-fruiting raspberries under Oregon conditions. *Proc. Am. Soc. Horticult. Sci.* **39**, 274–278.

Waldo, G. F. and Darrow, G. M. (1948). Origin of the Logan and the Mammoth blackberries. *J. Hered.* **39**, 98–107.

Walsh, C. S., Popenoe, J. and Solomos, T. (1983). Thornless blackberry is a climacteric fruit. *HortScience* **18**, 482–483.

Warmund, M. R., George, M. F. and Clark, J. R. (1986). Bud mortality and phloem injury of six blackberry cultivars subjected to low temperatures. *Fruit Var. J.* **40**, 144–146.

Watson, W. C. R. (1958). "Handbook of the Rubi of Great Britain and Ireland". Cambridge University Press, 273pp.

Way, D. W. (1967). A comparison of thorned and thornless clones of Loganberry. *In* "Report of East Malling Research Station for 1966", pp. 116–117.

Wenzel, W. G. and Smith, C. W. J. (1975). Germination tests with blackberry seeds. *Ang. Botan.* **49**, 11–14.

Williams, C. F. (1950). Influence of parentage in species hybridization of raspberries. *Proc. Am. Soc. Horticult. Sci.* **56**, 149–156.

Williams, C. F. and Darrow, G. M. (1940). The trailing raspberry—*Rubus parvifolius* L. Characteristics and breeding. "Technical Bulletin of North Carolina Experiment Station", No. 65, 13pp.

Williams, C. F., Smith, B. W. and Darrow, G. M. (1949). A Pan-American blackberry hybrid. Hybrids between the Andean blackberry and American varieties. *J. Hered.* **40**, 261–265.

Williams, I. H. (1959a). Effects of environment on *Rubus idaeus* L. II. Field observations on the variety Malling Promise. *J. Horticult. Sci.* **34**, 170–175.

Williams, I. H. (1959b). Effects of environment on *Rubus idaeus* L. III. Growth and dormancy of young shoots. *J. Horticult. Sci.* **34**, 210–218.

Williams, I. H. (1959c). Effects of environment on *Rubus idaeus* L. IV. Flower initiation and development of the inflorescence. *J. Horticult. Sci.* **34**, 219–228.

Williams, I. H. (1960). Effects of environment on *Rubus idaeus* L. V. Dormancy and flowering of the mature shoot. *J. Horticult. Sci.* **35**, 214–220.

Williams, W. (1957). Studies on *Rubus* species. *In* "Report of the John Innes Horticultural Institute for 1956", pp. 10–11.

Williamson, B. (1984). Polyderm, a barrier to infection of red raspberry buds by *Didymella applanata* and *Botrytis cinerea. Ann. Botany* **53**, 83–89.

Williamson, B. (1987). Effect of fenitrothion and benomyl sprays on raspberry cane midge (*Resseliella theobaldi*) and midge blight, with particular reference to *Leptosphaeria coniothyrium* in the disease complex. *J. Horticult. Sci.* **62**, 171–175.

Williamson, B. and Dale, A. (1983). Effects of spur blight (*Didymella applanata*) and premature defoliation on axillary buds and lateral shoots of red raspberry. *Ann. Appl. Biol.* **103**, 401–409.

Williamson, B. and Hargreaves, A. J. (1978). Cane blight (*Leptosphaeria coniothyrium*) in mechanically harvested red raspberry (*Rubus idaeus*). *Ann. Appl. Biol.* **88**, 37–43.

Williamson, B. and Hargreaves, A. J. (1979). Fungi on red raspberry from lesions associated with feeding wounds of cane midge (*Resseliella theobaldi*). *Ann. Appl. Biol.* **91**, 303–307.

Williamson, B. and Jennings, D. L. (1986). Common resistance in red raspberry to *Botrytis cinerea* and *Didymella applanata*, two pathogens occupying the same ecological niche. *Ann. Appl. Biol.* **109**, 581–593.

Williamson, B. and Ramsay, A. M. (1984). Effects of straddle-harvester design on cane blight (*Leptosphaeria coniothyrium*) of red raspberry. *Ann. Appl. Biol.* **105**, 177–184.

Williamson, B., Lawson, H. M., Woodford, J. A. T., Hargreaves, A. J., Wiseman, J. S. and Gordon, S. C. (1979). Vigour control, an integrated approach to cane, pest and disease management in red raspberry (*Rubus idaeus*). *Ann. Appl. Biol.* **92**, 359–368.

Williamson, B., Bristow, P. R. and Seemüller, E. (1986). Factors affecting the development of cane blight (*Leptosphaeria coniothyrium*) on red raspberries in Washington, Scotland and Germany. *Ann. Appl. Biol.* **108**, 33–42.

Williamson, B., McNicol, R. J. and Dolan, A. (1987). The effect of inoculating flowers and developing fruits with *Botrytis cinerea* on post-harvest grey mould of red raspberry. *Ann. Appl. Biol.* **111**, 285–294.

Wood, C. A. and Robertson, M. (1957). Observations on the fruiting habit of the red raspberry (*Rubus idaeus* L.) and on an occurrence of cane "dieback" in Scotland. *J. Horticult. Sci.* **32**, 172–183.

Woodford, J. A. T. and Gordon, S. C. (1978). The history and distribution of raspberry cane midge (*Resseliella theobaldi* (Barnes) = *Thomasiniana theobaldi* Barnes), a new pest in Scotland. *Horticult. Res.* **17**, 87–97.

Yarnell, S. H. (1936). Chromosome behaviour in blackberry–raspberry hybrids. *J. Agricult. Res.* **52**, 385–396.

Zurowski, C. L., Copeman, R. J. and Daubeny, H. A. (1985). Relative susceptibility of red raspberry clones to crown gall. *Phytopathology* **75**, 1289.

Zych, C. C. (1965). Incompatibility in crosses of red, black and purple raspberries. *Proc. Am. Soc. Horticult. Sci.* **86**, 307–312.

Zych, C. C., Hull, J. W. and McDaniel, J. C. (1967). Thornless erect blackberries may be bred from a wild selection. *Illinois Res.* **9**, 5.

Zych, C. C., Otterbacher, A. G. and Chu, M. C. (1968). Pollen pore number and polyploidy in *Rubus*. *Trans. Illinois St. Acad. Sci.* **61**, 421–424.

Subject Index

Index of Cultivars and Hybrids

(see also Appendix 1)